Kentucky Heirloom Seeds

Kentucky HEIRLOOM SEEDS

Growing, Eating, Saving

BILL BEST
with
DOBREE ADAMS

Foreword by A. Gwynn Henderson
Afterword by Brook Elliott

 UNIVERSITY PRESS OF KENTUCKY

Scholarly publisher for the Commonwealth,
serving Bellarmine University, Berea College, Centre College of Kentucky,
Eastern Kentucky University, The Filson Historical Society, Georgetown
College, Kentucky Historical Society, Kentucky State University, Morehead
State University, Murray State University, Northern Kentucky University,
Spalding University, Transylvania University, University of Kentucky,
University of Louisville, and Western Kentucky University.
All rights reserved.

Editorial and Sales Offices: The University Press of Kentucky
663 South Limestone Street, Lexington, Kentucky 40508-4008
www.kentuckypress.com

Unless otherwise noted, photographs are by Bill Best.

Library of Congress Cataloging-in-Publication Data

Names: Best, Billy F., 1935– author. | Adams, Dobree, author.
Title: Kentucky heirloom seeds : growing, eating, saving / Bill Best with
   Dobree Adams ; foreword by A. Gwynn Henderson ; afterword by Brook
Elliott.
Description: Lexington, Kentucky : University Press of Kentucky, [2017] |
   Includes index.
Identifiers: LCCN 2016058497| ISBN 9780813168876 (hardcover : alk.
paper) |
   ISBN 9780813168883 (pdf) | ISBN 9780813168890 (epub)
Subjects: LCSH: Seeds—Kentucky. | Seeds—Storage—Kentucky. |
Heirloom varieties (Plants)—Kentucky.
Classification: LCC SB118.38 .B468 2017 | DDC 631.5/2109769—dc23
LC record available at https://lccn.loc.gov/2016058497

ISBN 978-0-8131-8374-9 (pbk. : alk. paper)

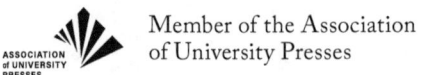

With special thanks to our sponsors:

Berea College

The Loyal Jones Appalachian Center

Tradition. Diversity. Change.

For
Vinson Watts (1930–2008)
&
The Vinson Watts Tomato

Contents

Color illustrations follow page 116

Carbonized archaeological specimens of the common bean (*Phaselous vulgaris*) assigned to varieties by Bill Best: top row, regular beans; second row, cut-short beans; third row, fall or October beans; bottom row, grits beans. (Hayward Wilkirson, University of Kentucky)

Foreword

Seed Saving:
An Ancient Kentucky Tradition

When I started doing archaeology in Kentucky in 1977, many residents told me Native peoples had never lived in Kentucky permanently but only hunted here. Undeterred, I moved to the commonwealth anyway and have been involved in researching the fascinating history of Kentucky's ancient Indian cultures ever since.

Over the years, I have seen the perpetuation of this falsehood—what I refer to as the Myth of the Dark and Bloody Ground—set up roadblocks to scientific understanding of Kentucky's natural history, justify stereotypes and myths that sustain bigotry about Native peoples, and deny Kentuckians an awareness of and appreciation for their state's role in the ancient world's agricultural revolution.

For these reasons, since the early 1980s I have made it my responsibility to dispel this myth and have pledged to speak to anyone anywhere in Kentucky about its ancient past and the peoples who once called this place "home." To share what I have learned about their diverse ways of life and their accomplishments. To tell, in my own way, their stories.

My area of expertise is not the study of ancient plants and

how people used them; that's what paleoethnobotanists and archaeobotanists do. I research the pottery made by ancient peoples in central and eastern Kentucky. But every archaeologist who has worked in a region for a long time learns the general outline of its history and knows and understands the important events and discoveries that have taken place there. I feel that the past holds much relevance—for us today and for folks in the future. Bill Best thinks so too.

When Bill and I met in January 2005, neither of us could have anticipated the surge in the "Buy Local" movement, the explosion of farmers' markets, or how relevant his heirloom seeds and seed-saving message would become. When we met, it was all about discovery on a personal level. I was struck by the fact that this man didn't need convincing about the truth of Kentucky's long Native history. Bill was struck by an image of eighteen charred archaeological beans I showed during my presentation (most are pictured at the beginning of this foreword). My colleagues and I had recovered them during our investigations at a 700-year-old Native farming village in north-central Kentucky. His response was that of a man whose decades-long hunch had finally been confirmed. There were so many similarities to beans he knew! They were cut-short beans!

I was blown away; he was blown away. Coming from two separate and very different places, it was a perfect harmonic convergence—a single lightbulb went on over both our heads. Those beans saved by generations of mountain people, who count Indians as part of their biological and cultural heritage, may be the living link, a tangible connection to Kentucky's ancient past. Native people never lived permanently in Kentucky? Not true! Central and eastern Kentucky's seed-saving roots go way back, more than 3,000 years. Western and southern Kentucky's seed-saving roots are just as deep, but that's another story.

Beans are newcomers to a very ancient Native plant food inventory. Current scholarship places the common bean (not its wild cousins and relatives) in Kentucky perhaps as early as 1,000 years ago, but certainly by around 800 years ago. But in truth, Kentucky's ancient peoples were seed savers and plant growers *millennia* before beans were even a bulge in a Native trader's pouch.

In the Beginning

Native peoples had been plant *gatherers* (for food, clothing, shelter, containers, dyes, fibers, herbs, and medicines) and animal hunters from the time they first arrived in Kentucky around 11,500 years ago. Down through hundreds of generations, they had gone from newcomers to inhabitants—from foreigners in a strange landscape to indigenous people living in an ancestral home.

Around 3,500 to 4,000 years ago, they started saving and planting the seeds of certain plants that grew where they lived. Why? That's easy to explain. Growing food gave them a measure of control, of predictability—a kind of food security. Some of these crop plants also were the source of raw materials needed for the manufacture of items used in domestic tasks, such as containers for storing food or water.

Other whys are not so easy to answer. No one knows exactly why early Native gardeners selected these *particular plants* from the many plants growing in their homelands. Were they tastier? Easier to prepare as food? More abundant? Higher in nutrition? More responsive to human intervention? Were they grown to be traded—the ancient equivalent of a seed swap?

We also don't know exactly why, after so many millennia of food gathering, indigenous peoples started saving seeds at that particular moment in time. Some paleoethnobotanists and archaeobotanists think climate and environmental changes likely had something to do with it. Or, perhaps to meddle with things

Nut gathering was a common and important subsistence activity that required knowledge of nut-tree locales and the manufacture of containers to transport the nuts back home. Additional processing was needed to recover the tender and nutritious nut meats from inside the hard shells. This woman is cracking nuts on a nutting stone using a stone pestle. (Kentucky Heritage Council)

is to be human. In any event, these ancient peoples, the long-ago ancestors of modern Indian groups, began to save seeds from specific plants growing near where they lived. Often, these were places where their own activities had disturbed the land.

Growing food changed Native peoples' relationship to the natural environment. They were modifying it in a way that hunting did not. Growing food also changed their relationship to the spirit world. Over time, in the process of domesticating plants, people domesticated themselves. They began to build more permanent settlements and live in larger groups.

A World Hearth of Plant Domestication

The places where people first *domesticated* native plants (adapted wild plants for human use) are called *world hearths of plant domestication*. People living in these spots were the original seed savers. They were the first to select seeds with certain desirable traits, such as a larger size or a thinner seed coat (the protective outer layer), from particular wild plants and then to raise those plants by gardening and farming. Over time, these choices of which seeds to save domesticated the plants.

The five most familiar hearths—Mexico (corn), Peru (potatoes), the Middle East (wheat and barley), Africa (soybeans and millet), and Southeast Asia (rice)—are those where people first domesticated the foods we commonly eat today. A sixth hearth, eastern North America, is less well known. A key reason that we know about it is the research archaeologists have carried out at sites in eastern Kentucky's Red River Gorge. Sites there contain some of the oldest and best-preserved evidence of plant domestication in this region. That's because the dry conditions inside many rock shelters and the very ashy soils in some others—the ash acts to neutralize Kentucky's generally acidic soils—preserve the fragile charred and uncharred seeds.

Native groups domesticated "fleshy" squash, with its thin skin and soft tissue. They also domesticated or *cultivated* (grew the plants but did not change them physically) seven "weedy" annuals. They domesticated sunflower, marsh elder, and goosefoot, and they cultivated maygrass, erect knotweed, giant ragweed, and little barley. These plants produce nutritious seeds that are good sources of oils and fats (sunflower and marsh elder) or starchy carbohydrates (goosefoot and the rest). These annuals thrive in drier places in river floodplains and forest clearings where either natural or human activities have recently disturbed the ground. They produce seeds in the spring through early summer (maygrass and little barley) or in the late summer and fall (sunflower and the rest).

For two millennia, between 3,500 and 1,000 years ago, eastern Kentucky's indigenous peoples saved the seeds of native plants and grew them. Because it was so long ago, we do not know exactly how they planted, tended, and harvested their crops, but researchers have some ideas. It seems likely that these peoples planted their seeds in garden plots—small openings on hill slopes and in floodplains in full sun. They maintained their gardens with fire, burning off early weeds and the previous year's dead growth in the late spring. Then they planted their seeds in moist soil, possibly as monocrops, either broadcasting the seeds or planting them individually, depending on the species. In time, these early seed savers became hunter-gatherer-gardeners.

Squash, the Earliest

Fleshy or pepo squashes (*Cucurbita pepo*) include orange jack-o'-lantern pumpkins; acorn, zucchini, and pattypan squashes; and numerous types of ornamental gourds. Squash is a good source of fiber, vitamins, minerals, and folic acid. Its seeds are good sources of oil and protein.

People first domesticated pepo squashes native to Mexico around 10,000 years ago. Several millennia later, around 5,000 years ago, people living in eastern North America independently domesticated pepo squashes native to their homelands. This process began when local hunter-gatherers collected and ate the protein-rich seeds of small, thin-skinned wild gourds. As time passed, they carefully selected certain wild gourds that best met their needs, saved the seeds, and planted them. This changed the plants physically: fruits became larger and fleshier, rinds became thicker, and seeds became larger too.

A little over 1,000 years later, Native hunter-gatherer-gardeners were growing smooth, lobed, and warty varieties of squash. A rock shelter in eastern Kentucky has produced some of the oldest pepo squash seed and rind fragments in eastern North America, dating to about 3,700 years ago. By about 1,400 years ago, Native peoples' food needs had changed again. Their efforts to breed edible, meaty squash produced thick-fleshed fruit with thin rinds.

Native peoples planted squash in late spring along the edges of garden plots where the vines had room to spread or climb. They harvested the ripe fruits in the summer. They likely ate the seeds and flowers as well as the fruits. They would have eaten the seeds whole or ground them up to add to stews, taking care to dry and save some for planting the next spring. They could have roasted or baked the fruits whole in hot coals or sliced them into strips to dry for later use. It is also likely that Native peoples dried the fruits whole to use as durable containers, rattles, or bobbers and floats for fishnets.

Domesticated Weedy Annuals

Oily-seeded, multiheaded sunflower (*Helianthus* spp.) was an early food crop domesticated in ancient Kentucky. From year to year, indigenous gardeners selected the largest sunflower seeds to plant.

For this reason, domesticated sunflower seeds are longer and bigger than their wild cousins. They are a good source of healthy oils and fats as well as important minerals and vitamins, protein, and fiber. The oldest domesticated sunflower seeds in Kentucky date to about 3,100 years ago.

Starchy-seeded goosefoot or lambsquarters (*Chenopodium* spp.) is a bushy plant with fleshy stems. The leaves are shaped like a goose's foot, and each plant produces dozens of seed clusters with thousands of tiny seeds. These grains have more minerals and certain amino acids than other cereal grains, and they lack gluten, a common allergen in wheat.

As early as 8,500 years ago, Native peoples in Kentucky collected wild forms of goosefoot for its nutritious shoots, leaves, and seeds. But beginning about 3,700 years ago, they began to grow goosefoot, which caused physical changes in the plant. Compared with wild forms, the seeds of domesticated goosefoot are larger and the seed coat is thinner. Domesticated goosefoot seeds from eastern Kentucky rock shelters have been dated to 3,500 years ago, making them among the oldest found in eastern North America.

At first, sunflower seeds were likely to be sown by broadcasting them, but as they became larger, Native gardeners likely switched to planting sunflowers in hills along garden edges. Spaced about one foot apart, the plants could grow to over eight feet tall. Small-grained goosefoot was probably planted by broadcasting the seeds, either alone or intermingled with those of another plant like knotweed, which shared goosefoot's growth habits, ripening schedule, and harvesting and cooking requirements. This method is called "grown as maslins." These plants grew densely, shading out weeds. Since goosefoot sets more seeds when the plants are not overcrowded, Native gardeners likely thinned the stands, eating the young plants as greens.

Ripe sunflower seed heads were ready for harvest in late

summer and fall, which was easily accomplished by handpicking or cutting the heads from the stalks. Goosefoot seeds ripened in the late fall, typically in October. Native gardeners harvested those seeds by hitting uprooted or standing plants, using either their hands or a stick, and catching the seeds in baskets or on hides or blankets. Alternatively, they could have stripped the seeds from the stalks by hand. Seeds of both plants were easily stored in baskets, bags, and subterranean pits for use year-round and for planting in the spring.

Native peoples sometimes ate the entire sunflower seed, shell and all, or else they ate just the meat inside the shell, as we do today, alone or combined with other seeds in a trail mix. Goosefoot grains were threshed, winnowed, and parched to remove the chaff. Native cooks ground up both sunflower and goosefoot seeds by drawing handheld stone tools back and forth across flat grinding stones or by pounding the seeds with wooden or stone tools in man-made depressions in large boulders, called hominy holes or bedrock mortars. Finely ground goosefoot grains were made into flour. Coarsely ground sunflower and goosefoot seeds were mixed with meats and other seeds to make stews and gruels cooked in clay pots. In the spring and summer Native peoples also ate tender goosefoot leaves as greens, either raw or cooked. Because of goosefoot's medicinal properties, Native herbalists used young shoots to treat fever, intestinal parasites, and bruises. Archaeologists have found goosefoot seeds at domestic sites and at burial mounds and sites where Indian peoples prepared their dead for burial. At these ritually charges places, mourners may have eaten the plant during ritual feasts.

The Three Sisters

But all things come to an end. About 1,000 years ago, the hunting-gathering-gardening peoples of central and eastern Kentucky stopped saving, growing, and eating many of the weedy annuals that had been such an important part of their gardening heritage. People were turning with a passion to farming, and seed saving, planting, and growing began to occur on a much grander, larger, and more intensive scale. Pepo squash, sunflower, and perhaps goosefoot remained in their seed inventory, but the other plants of their ancestors reverted to weeds. New seed crops took their place.

Diffusion is the process by which people accept plants domesticated elsewhere into their own food-growing tradition. People share the seeds as well as the knowledge needed to grow the plants and prepare the foods. A foreign plant will thrive in a new environment if it meets the plant's growing requirements or if people can adjust the local environment to meet the plant's needs. In this way, people continue to change and manipulate plants, producing new varieties over time.

Through the process of diffusion, seeds originally domesticated in hearths in Mexico and South America appeared in Kentucky for the first time. After a thousand years, Kentucky's Native inhabitants, like groups all across North America, were mainly planting, growing, and eating three varieties of corn. A few centuries later, beans arrived. Saving the seeds of these foreign plants became part of Native lives.

Corn or maize (*Zea mays*) was domesticated between 6,000 and 9,000 years ago in Mexico. It diffused across several thousand miles to Kentucky via the American Southwest in two waves separated by several hundred years.

The earliest traces of corn in Kentucky are charred kernels found at a site in central Kentucky that date to about 1,400 years ago. But if ancient pollen grains are any indication, corn likely arrived in Kentucky much earlier, by around 1,800 years ago. Kentucky's Native peoples did not immediately grow corn for food—that took between six and eight centuries. Before it became

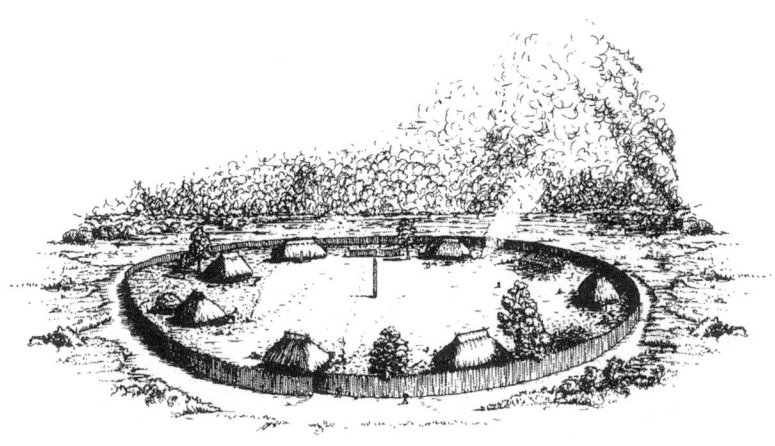

Palisades made from small trees chosen for their straight trunks and decay-resistant wood sometimes surrounded hunter-gatherer-farmer villages. Ceremonies, dances, or just plain socializing took place in the central plaza. (Kentucky Heritage Council)

a major foodstuff, archaeologists think corn played an important role in ritual and ceremony.

Growing corn changed time-honored planting and harvesting practices, foodways, and lifeways. Villages with houses arranged around a central plaza were scattered all across central and eastern Kentucky beginning about 800 years ago. Fields of corn, beans, and squash surrounded each village.

Planting became more labor intensive, changing from broadcast sowing to carefully planting seeds in hills using a dibble stick. Harvesting changed too: from uprooting or hand-stripping seeds and threshing them to "plucking and shucking." In central and eastern Kentucky, Native hunter-gatherer-farmers grew mainly Eastern 8-Row flint corn. Its kernels, typically large and crescent shaped, are arranged in eight rows around the cob.

The common bean (*Phaseolus vulgaris*) was domesticated in the Peruvian Andes by at least 8,000 years ago and later independently in Mexico by at least 6,000 years ago. Archaeologists are not exactly sure what route the plant took to get to Kentucky. Some think beans came from the American Southwest, spreading rapidly across the Great Plains and into the Midwest and Northeast. Others think beans diffused northward from the Caribbean, across Florida, up the Atlantic coast, and then east into Kentucky. Regardless of the path, it is quite clear that domesticated beans were the last of the major cultivated plants to appear in the Native plant food inventory.

Native peoples adopted beans quickly, over the span of a little more than a century. By about 725 years ago, Native groups were growing beans across much of eastern North America. The earliest domesticated Kentucky examples were recovered from sites in Jessamine and Fayette Counties dating from perhaps 1,000 years ago, but certainly by around 800 years ago.

The Native planting system—intercropping, or planting complementary crops together, where space was limited, or monocropping, where space was not—consisted of planting in hills spaced in a regular pattern or in fields. Sophisticated, sustainable, and productive, this system depended on the same kind of knowledge about crops, soil management, and fire use as farming does today.

The practice of pairing a cereal grain with a legume is found in almost every agricultural community worldwide and throughout time. In central and eastern Kentucky, the cereal grain was Eastern 8-Row flint corn and the legume was beans. When corn and beans were added to squash, the ancient and familiar standby, it resulted in what many eastern North American Native peoples still refer to as the "Three Sisters" planting system.

In the spring, Native men prepared the fields by burning off the remains of the previous year's harvest. Using dibble sticks, women planted corn singly in rows or in small groups in hills, along with groups of bean and squash seeds. Given its productivity, corn was the engine of the system. Stalks served as the physical support for the bean vines. Beans replenished nitrogen in the soil, serving as a kind of natural fertilizer that made the other plants grow better. Squash reduced weed pressure and, by shading the ground, helped retain soil moisture and decrease soil erosion. This system required periodic weeding and pulling up or *hilling* the soil around the plants to promote root development. Women used hoes for this task, made out of thick freshwater mussel shells or elk or deer shoulder blades lashed to wooden handles. Given corn's attractiveness to animals, Native farmers had to be especially vigilant as the time for harvest drew near.

Lacking large domesticated animals as a source of protein, Kentucky's Native peoples depended on wild animals such as deer, bear, elk, and turkey and on vegetable protein to meet their nutritional needs. Dishes that combine corn and beans are a good source of complete vegetable protein. Beans have the highest protein content within the vegetable kingdom; they are a great source of fiber, are rich in iron, and contain lysine, a vital amino acid. Corn is high in calories and carbohydrates but is relatively low in protein, and it lacks important amino acids. Squash contributes calories, dietary fat, vitamins, and minerals not found in corn or beans, and its seeds are good sources of protein and oil.

After harvesting in the fall, the women plucked or cut the ripe ears of corn off the stalks. In central and eastern Kentucky, Native peoples stored corn in cribs above the ground or hung the cobs inside their homes, the husks braided together. They prepared corn in many different ways. They picked it green and roasted the ears in coals. Corn scraped off the cob was made into soups or gruels cooked in large jars and served in bowls. It was made into fried, baked, or boiled breads. They parched corn, pounding it into a meal they used when traveling. Letting dried corn ker-

Kentucky Native farmers made jars like these from local clays. There were no potters' wheels; they built the pots from coils of clay. Jars were often decorated with geometric designs etched into the necks. They were made watertight by fire-hardening in fires built on the open ground. (Kentucky Heritage Council)

nels sit for several days in water mixed with wood ash removed the hull and produced soft and tender hominy. To harvest beans, they picked the pods from the vines or allowed the vines to dry before pulling the pods and threshing them in the fields. Either in the pod or shelled, beans were cooked in soups and stews. A portion of every harvest was stored for planting the next year. Saving seeds in clay vessels for spring planting protected the precious inventory from rodents and moisture.

Though a newcomer, corn became deeply intertwined with many aspects of Native life. Corn husks were used in the manufacture of masks, moccasins, mats, and baskets; spent cobs were used for fuel. In eastern North America, corn was intimately tied to ceremony and religious practices and beliefs. The Green Corn Ceremony, held in late summer, was and remains a celebration of thanksgiving, of prayers and feasting, of dancing. For some groups, this ceremony was also associated with annual cleansing and purification of the individual and the community.

Unlike some of the other plants they grew, beans and bean pods did not furnish the raw material for containers or masks, as gourds did, or for mats and baskets, as corn husks did. Nor were beans celebrated in ceremony in the same way as corn. However, archaeologists have discovered that for centuries, shelled beans were part of Native burial rituals. Along with shelled corn, beans were given as offerings or eaten as part of mourning rituals or burial ceremonies or feasts. At sites in Mason and Bourbon Counties, archaeologists have found concentrations of charred, shelled beans and corn that may represent grave offerings. It was the photograph of such charred beans recovered from this kind of context that Bill identified as cut-short beans when we first met.

The Native Gift

The earliest European explorers described Native corn-bean-squash fields in eastern North America. Not long after the Europeans arrived, diseases of European origin "diffused" from group to group, devastating these Native hunter-gatherer-farming cultures.

European American colonists and their slaves learned about corn, beans, squash, sunflowers, and a host of other indigenous plants from the Indians. Pioneers brought the seeds of these plants, as well as seeds of more familiar Old World grains such as wheat and barley, with them to frontier Kentucky. But Kentucky's climate and weather conditions were not optimal for the pioneers' familiar crops, and its Bluegrass soils were too fertile at first. So pioneer farmers sowed the unfamiliar corn in roughly cleared fields. As it was for their Indian contemporaries, corn was the dominant crop on the Kentucky frontier and a dietary mainstay.

For a time, Kentucky's Euro-American farmers even modeled their growing practices on those of Native farmers, multi-cropping corn with beans in hills. Documents describe how pioneers planted their fields "in check." First they plowed four-foot furrows one way, and then they plowed four-foot furrows perpendicular to them, planting seeds at the intersections. Later they hoed the intersections into hills. Like Native peoples, they too prepared corn as hominy and cornbread and saved the seeds.

In Kentucky's early statehood period, farmers took their corn harvest to mills, where it was ground into flour and meal. They fed field corn to their livestock. They turned their corn into bourbon whiskey and moonshine. As time passed, farmers were just as likely to plant Old World grains (wheat and barley) as they were the New World ones. For a time, wheat and barley eclipsed corn and beans as important crops. The Civil War changed that. Beans—cheap, nutritious, and filling—became the staple food for soldiers, whether they were fighting for the North or the South.

After the war, farmers grew corn, beans, and squash, among other crops. And they saved seeds.

Seed Saving Today

Today, people living in central and eastern Kentucky grow a wide variety of pepo squashes, sunflowers, sweet corn, and beans in backyard gardens. Most are grown from more recent hybrid seeds. The goosefoot found in Kentucky today is the wild variety. Twenty-first-century gardeners consider it a weed, and few appreciate its nutritional value and history, although some Kentuckians now eat a relative of goosefoot called quinoa (*Chenopodium quinoa*), domesticated in Peru at least 7,000 years ago.

Many eastern Kentuckians privilege their Native heritage as one strand in their rope of identity. And seed-saving traditions have endured for generations in some places in the state. The old varieties are now called Kentucky "heirlooms."

Planting and growing food crops, and selecting seeds and saving them to plant again, are part of a seasonal round that has gone on for millennia in Kentucky. Humans domesticated plants by the simple act of saving seeds, and over time, seed saving tied people more closely to the places where they planted and harvested their crops. In the process, humans domesticated themselves. We are inextricably linked to the plants whose seeds we save and sow again. Since the middle of the twentieth century, many people have forgotten this strong connection.

Bill Best rightly makes the case that seed saving and plant-food growing in Kentucky have deep historical roots. If not for his and his colleagues' interest, we might have lost this connection to our ancient heritage. Bill and his seeds remind us that saving seeds is a Kentucky tradition.

A. Gwynn Henderson

The Vinson Watts tomato. (Dobree Adams)

Preface

Seed saving is nothing new in Kentucky. It has been going on for at least two millennia; the quality and quantity of seeds saved meant the difference between dying and living to plant again another season.

As Gwynn Henderson tells us in the foreword, seeds were saved for thousands of years by the Native Americans living in the area of real estate that we now identify as Kentucky. Parts of the state had been farmed intensively for multiple generations. These Native Americans freely shared their seeds with European settlers and often intermarried with them. The seeds they shared, especially corn, beans, and winter squashes, became the basis for the diets of generations of Kentuckians. In his afterword, Brook Elliott (who was thrust into seed saving when he found that 80 percent of eighteenth-century vegetable varieties were extinct) debunks several myths about the food found on colonial tables.

The saving of seeds came to be an important pursuit. As late as the 1950s, seed saving was one of the most valuable skills for young people to learn, and it was passed down from grandparents to parents to children in a seamless fashion. The first seed catalogs, from small, mostly family-oriented companies, sold seeds for vegetables that had been in these families for many years. Quality was not compromised by producing seeds for large corporate and highly mechanized farms. One could still grow and eat, with some degree of satisfaction, the fruits and vegetables grown from these first commercial seeds.

This situation was too good to last. Multinational food, feed, chemical, and fertilizer companies started buying up seed companies and producing seeds primarily suited for mechanical harvest, long-distance transportation, and a long shelf life in large grocery stores. Thousands of varieties of excellent-quality fruits and vegetables were discarded like so much junk.

At the same time, gardening was falling out of favor. Those serious gardeners who complained about the poor quality of seeds were like so many voices in the wilderness. Land-grant universities assisted in producing hybrid seeds for new fruits and vegetables, leaving gardeners only one recourse: start saving seeds again and recover seeds that otherwise might be lost. *Kentucky Heirloom Seeds* is about that recovery in Kentucky.

1

Seed Saving

Past, Present, and Future

After fifty-plus years of collecting, growing, selling, and sharing seeds of heirloom beans, tomatoes, and a few other vegetables, I think I can generalize about the types of people who save seeds. I use the term *traditional seed savers* to refer to individuals who have never been separated from the land and who are following in family pathways.

In July 1988 the *Rural Kentuckian* (now *Kentucky Living*) published a well-written article by Judy Sizemore about my family's farming operation that focused on our work with heirloom beans and tomatoes. Within five days we started getting letters and phone calls from people throughout Kentucky. Some wanted to get seeds for the greasy beans pictured in the article, while others wanted to send us some of their seeds. A man from Laurel County wanted to trade Nickell beans he had gotten years earlier in Elliott County for our greasy beans. Ruth Thomas sent me some of her Paterge (Partridge) Head beans from Clinton County. Her son Rudy had worked for me several summers in the Upward Bound program at Berea College.

And that was just the beginning. Within six months we had received letters from eighty-six people asking about beans, wanting to trade beans, and sharing bits and pieces of historical information.

Tomatoes ready for eating! The Willard Wynn Yellow German tomato, one of our heirloom favorites. (Dobree Adams)

A few wrote about tomatoes, but most wanted to talk about beans. Some even lectured me a little. Others contacted us by phone or paid personal visits to our farm. And all this happened in the days before the Internet!

Even more astounding: some of the letter writers lived outside Kentucky (they came from six different states). Most of them were native Kentuckians who had migrated elsewhere to find work but were still gardening and swapping seeds and growing information. A few letters came two years later, referring to the same article, and to this day I occasionally run across people who have held on to that article.

Traditional Seed Savers

Some old-time seed savers can't read and write, but they are intellectually sharp and know the value of traditional forms of knowledge. One of them was my next-door neighbor for many years, who taught me a lot about saving and growing heirloom vegetables. These old-timers know that seed companies have been putting out inferior products, although they might not be able to figure out why. Many are high school dropouts who have gardened all their lives and remember the hard times of the Great Depression. They have maintained the survival skills they grew up with and have no intention of becoming dependent on the benevolence of large corporations and giant factory farms. Some of them are retired factory workers and miners who are now dealing with crippling disabilities or respiratory problems, but they garden as if their lives depended on it. Most like the feeling of having a certain degree of independence.

Others are better educated but want to maintain their survival skills outside the system, either because they want to be prepared for the worst economically or because they have a simple love of growing things. I often hear them say, "I just like to see things grow." Some are college graduates with several degrees, but they never "got above their raising" and strive to keep in touch with the ways of their ancestors and maintain historic ties with the land. I never cease to be amazed at the sense of history of so many seed savers, regardless of the amount of formal education. Many talk about great-uncles or great-grandfathers who fought on one side or the other during the Civil War. Border-state Kentuckians fought on both sides—often brother against brother or father against son. Or they speak of great-great-aunts or great-great-grandmothers who were renowned seed savers and had particular varieties named after them.

The Lexington Farmers' Market

The farmers' market in Lexington is one of Kentucky's oldest continuous farmers' markets in modern times. We started it in 1973 and, despite numerous setbacks, it has grown larger almost every year since then. Like me, a few of the founders had farmed and gardened all their lives, but several were going "back to the land" and saw the new market as a way to enter farming quickly and maybe even make a quick buck. Demand was heavy the first summer, with most vendors selling out by midmorning.

Local newspapers gave the market good coverage, and it didn't take long for displaced eastern Kentuckians to start coming in droves to buy beans, tomatoes, potatoes, onions, sweet corn, greens, cantaloupes, watermelons, and all the other fruits and vegetables they had left behind in their own gardens when they migrated to Lexington to find work. People from other countries living in Lexington quickly discovered the market, and okra was especially popular among them; they usually bought all the okra available during the first few minutes of the market's opening. Many people native to the Bluegrass had never eaten okra, but a number of vendors started growing it the next year to meet the high demand, especially from foreigners and southerners.

I am now the sole member of the original group still selling at the market. I started out as the youngest founding member and am now the oldest grower. I began selling heirloom beans soon after the market opened, growing what many people call "real" beans because

Mrs. Elizabeth Tompkins has been patronizing the Lexington Farmers' Market for decades and is probably my oldest customer. She celebrated her 100th birthday in September 2015. Here she is holding one of my tomatoes and one of my books. She will probably be first in line to buy this book!

I allowed them to become "full" prior to picking. I was simply growing the beans I liked, and I found that many of the customers liked them too.

Most of the other vendors at the market were growing and selling, or buying and reselling, bush beans, primarily stringless modern varieties such as Blue Lake and Tenderette. They were charging about two-thirds the price I was asking for my beans, so a lot of customers, being thrifty, bought the cheaper varieties, even though they didn't look like the beans they had known in the past. It was sort of like buying a Yugo automobile—cheap, but not of very good quality.

I soon noticed that many customers would come by my stand and purchase a few of my beans to "flavor" the cheaper ones they had just bought. When I asked them to explain, they said they wanted to have some shelled beans to go with the others, which were just hulls. When my beans were cooked, the beans inside the hulls broke free, giving customers at least a hint of the beans they had grown themselves before moving to Lexington, where they often didn't have enough space to plant a garden.

Once it was apparent that the farmers' market was going to become a fixture, people from eastern Kentucky and from the mountain regions of other Appalachian states, where gardening was still practiced on a large scale, began to bring their family bean seeds to the market. They would give them to me, Ott McMaine, and a few others who were willing to grow them. And soon after that, customers started putting in orders for canning beans before the selling season began, a practice that continues to this day, more than forty years later. Several other growers got the hint and began growing the older types of string beans—beans that are now called heirlooms.

In the meantime, sellers of commercial beans were charging higher prices for machine-picked beans, which made heirloom beans even more attractive. There were no longer any cheap beans at the market, so there was no reason for customers who favored the heirlooms to mix them with commercial beans.

Concurrent with this renewed interest in "old-fashioned" or heirloom beans, heirloom tomatoes were also making a comeback. When the seed companies started selling mostly hard commercial-type tomatoes to gardeners, a slow rebellion started to emerge against tough and tasteless tomatoes, much akin to the rebellion against tough beans. Small specialty seed companies started to offer heir-

Growing early tomatoes in the high tunnel. (Dobree Adams)

loom tomato seeds, and farmers' market growers began to grow and sell the older varieties of tomatoes and save their seeds.

Farmers' market customers, some of whom also ate regularly at restaurants, suggested to restaurant owners and chefs that they look into buying heirloom tomatoes for their recipes, and the demand for heirloom tomatoes took another big leap. Restaurant owners, who had previously bought only hybrid, commercial-type tomatoes from large food distributors, suddenly started buying only heirloom tomatoes from farmers' markets because of their freshness and superior flavor and texture. They featured heirloom tomato dishes, sometimes listing the names of the heirloom tomatoes, as well as the growers, on their menus.

As a rebellion against "fast food," an international movement started in Italy in 1986. Called "Slow Food," this movement brought additional attention to older types of fruits and vegetables and other food products, and it encouraged people to think a little more about the foods they were eating, how they were prepared, and what modern foods were doing to their bodies.

Several widely publicized food recalls in the 1990s convinced an already nervous public that they should pay attention to the sources of their food as well as its nutrient content. E. coli, listeria, salmonella, botulinum, and other germs suddenly became of interest to ordinary people, not just to microbiologists.

A new word entered the language: *locavore* referred to an individual who ate locally grown foods. At the same time, a "Buy Local" movement was occurring in many places, sometimes promoted by state agriculture departments, to encourage more interest in buying foods grown locally and in season. With the environmental movement pushing the idea of lessening one's carbon footprint, buying local seemed to be an idea whose time had come. All this tended to make heirloom fruits and vegetables grown by local producers more attractive and offered a logical way to improve one's health and to be more environmentally responsible, not to mention supporting the local economy and helping small farmers diversify and survive.

Keepers of the Seeds

When I first met Gwynn Henderson, she was giving a presentation that included a big-screen photo of bean seeds found in Kentucky that were several hundred years old. I was astounded because they looked so much like the beans I knew, and I am still intrigued by those ancient beans. As Gwynn reveals in the foreword, by the time Europeans began settling in the southern Appalachians, Native Americans had already established their seed-saving traditions.

The earliest evidence I have is from a college friend whose paternal ancestor arrived in the New World at the time of the American Revolution. Drafted to fight on the losing side, after the war he was given the choice of going back to Scotland or moving into the moun-

The Cherokee Trail of Tears bean, a well-known heirloom of the southern Appalachians, is said to have been taken by the Cherokees on their forced march to Oklahoma. It has a beautiful fuchsia or maroon color.

tains of what would become western North Carolina. He chose the mountains, and soon thereafter he married an Indian woman. One of her contributions to the marriage was some greasy bean seeds. These beans are still part of the family's seed collection some 240 years later. Of necessity, settlers saved seeds, traded with one another and with the Indians, and were heavily dependent on beans as a vital part of their diets.

Harriette Arnow, in her splendid and well-researched book *Seedtime on the Cumberland,* spends a lot of time discussing gardening and dietary habits. Beans are mentioned on fifteen pages, but tomatoes do not even make the index. Tomatoes were not common in early Kentucky, but they came in later with a vengeance. Arnow's book talks about growing beans, preparing them for the table, and preserving them to be eaten during the months when gardens were not producing.

Almost every extended family in Kentucky, especially those from the eastern and south-central parts of the state, has one or more members who are growers of traditional beans and keepers of the family bean lore. Family reunions feature many bean dishes prepared in all the traditional ways—fresh green beans, fresh shelly beans, shuck beans, dry beans, pickled beans, and sulfured beans. And bean lovers usually sample all of them.

Some families still have bean stringings at canning time, when many bushels of beans have to be strung in a relatively short period in order to can many quarts of beans, sometimes numbering in the hundreds. Canned beans are often given as gifts to visiting relatives or to strangers who just happen by. Several gardeners I have interviewed insisted that I take home one or more jars of beans from the previous (or earlier) summer. Properly canned beans can last for years when stored in a dark, cool place.

Seed Saving and Culture

Traditionally, extended families saved enough seeds to get them through another year. In addition, community leaders such as preachers and politicians contributed to the dispersal of seeds to people outside the extended family. Preachers were often invited to eat with a family in the congregation after church, and they simply passed on the gift of beans received the previous Sunday to the next provider of dinner. Many of these bean varieties came to be called "Preacher Beans." There are also numerous stories of politicians carrying bean seeds with them when visiting potential voters; they traded seeds with the voter and then moved on to the next house down the road, completing the cycle once again. I don't know of any varieties called "Politician Beans," but the stories persist. Preachers were typically held in higher regard than politicians.

I witnessed such seed trading as a child. My mother carried on seed-swapping rituals right up to the end of her life (she died a month short of her eighty-third birthday). We always had good beans to eat, and pots of beans were almost always on the cookstove, ready to be warmed up by adding a few sticks of wood to the fire. The beans never got completely cold and became better each day as the cured pork seasoning permeated the remaining beans.

I remember picking beans with my mother when I was just a few years old. I was captivated by the colors of the Hill family bean and by the bright green of the packsaddle stinging worm (often found feeding on the underside of corn leaves during silking), which my mother taught me to avoid at all costs. About ten or twelve years ago my cousin Clarine Green Best sent me some of those Hill family beans with this note:

These are the beans that we found in Ben's mother's can house in an open half-gallon jar after she had been gone for 15 or 20 years. We planted them and they came up. She had told us they had been in her family since she was a small child and that she used to carry them in a little apron and drop seeds for her mother who would plant them in the cornfield. In the fall they would pick bushels of these to pickle and to dry. When dry, they threshed them out on a large tarp and used as dry beans.

It would have taken a lot to make a mess since the Hills had 12 children. Ben's mother, Cecile, would sit on her porch and fix these. She would shell out the ones that had gone to seed. She always threw away the black ones. Ronnie Hawkins remembers she used to let them have the black ones to play with. They would take bean hulls and make fences and pretend the shelled beans were cows. Of course, she had to let them have a few of the other colors, too, so they could have different colors of cows.

The first eight Hill children were all girls; the youngest four children were all boys. The girls learned early on how to plow and do all the other farmwork. One girl would guide the horse along the row to be plowed while another would handle the plow. They were an amazing family, and many of their descendants still live in Upper Crabtree, my home community in western North Carolina.

Ten years after Mother's death, my sister Janet, who had inherited the house where we were raised, called to ask me to come and check Mother's freezer, which still had some containers of seeds. I found thirteen varieties in her freezer, all in airtight containers. Some had never been opened, since Mother had been unable to do much gardening in the last two or three years of her life. One of those unopened containers had been given to her by my father's ninety-year-old first cousin. Called the Lazy Daisy bean, it is one of the best greasy beans in existence. The following summer I grew twelve of the varieties from her freezer, with nearly 100 percent germination. Seven years later, when Janet cleaned out the entire freezer, she found seven more packages of beans at the bottom, all of which were still good.

In Kentucky, Shakers are considered to be the earliest to save seeds and then sell them in packets to the general public. The nature of their settlements gave them the advantage of having enough labor to market their fruits and vegetables and their seeds. Many converts to the religion were widows or widowers with children who were disconnected from their biological extended families. Working in a

Jenny Sloan holding her mother's Minnie butter beans, stored safely in her freezer. Another old-timer, Jenny died recently.

communal way allowed all members to be involved with the farms and gardens and with selling animals, produce, and packaged seeds.

In the past, most traditional seed savers in Kentucky were women, who were primarily responsible for the family gardens. Most of the men who are seed savers today (and there are many) acknowledge that their mothers, grandmothers, aunts, and great-aunts taught them that seeds had to be saved. Few current seed savers were introduced to the practice by their fathers.

Based on my numerous interviews of traditional gardeners and seed savers, I believe that women were also largely responsible for noting "sports" in bean rows. These are the crosses or mutants that are earlier or later than the other beans or have a different shape, size, or color. Often these mutants were kept and planted the following year, separately from the other beans. This is how new varieties come about, and based on my own experiences of the past

Harvey and Ada Dicken, outstanding gardeners from Clinton County. Harvey, the nephew of Jenny Sloan, was a student in my first Upward Bound program at Berea in 1967. Harvey is Cherokee and Blackfoot; Ada is Cherokee.

fifty years, I'm convinced that most family beans came into existence by this route.

I'm further convinced, after decades of growing, selling, and observing heirloom beans as they grew, that within each bean lies the potential for hundreds or even thousands of other varieties. I have isolated several new varieties myself, some of which are outstanding beans and among my best sellers. I have also observed beans grown by others that have produced outstanding mutations, some of which I now grow myself.

Traveling Beans

Although I have never made a particular effort to trade bean seeds with customers at farmers' markets, I have received many gifts of bean seeds. During the past few years, I have insisted on gifting seeds back to those who give them to me, usually with success, since so many people now know the value of maintaining as many heirloom varieties of fruits and vegetables as possible.

I remember the first bean ever given to me at the Lexington market in the late 1970s. It was a brown greasy bean I obtained from a man traveling through Lexington on his way to visit his hometown in Harlan County. While living in Cincinnati, he had continued to grow his family beans from Harlan and had started experimenting with them. He told me how he selected seeds from the earlier blooming plants and gradually developed a strain of his family bean that matured two weeks earlier than it had in years past. Unfortunately, I didn't write down his name and address and didn't get any information about his family, since I had not yet started an organized effort to save old seeds and note their history. Although I knew the old varieties were far superior to the beans being bred by seed companies, I had not developed any sense of urgency. I was selling many bushels of beans each summer, but I was not saving seeds in any systematic fashion with proper documentation.

That year, there was still enough time left in the season to plant his beans and see how they performed for me. As the plants grew, I noticed that one was growing taller than the others, and when the plants started blooming, this one had a different color bloom—white—and larger leaves. As the beans matured, the plant with the white-colored blooms was producing a much larger bean than all the others. When the beans matured completely and were ready to pick, the odd vine had produced a long pod with a white-seeded bean. All the other beans bred true—short pods with brown seeds.

I had the good sense to save the apparent mutant and put the seeds in the freezer, but I waited several years to plant the beans to see if they bred true. When I finally decided I couldn't procrastinate any longer, I planted all sixty-one seeds I had saved from that one plant. We had a heavy rain shortly after the seeds were planted, and only nineteen broke through the ground and survived. However, I got enough beans to eat for one meal and then share a few with other people to grow the following year. The beans were quite tender and tasty, and they were all of the same type as planted. The bean was breeding true. Since there were no other beans planted nearby, I tend to think it was a mutant.

This was my first bean discovery, so I initially called it Bill's Original. Then I decided it wasn't appropriate to name it after myself, so I named it the Robe Mountain bean, after the mountain behind my house. We have since grown and sold beans and seeds by that name. Interestingly, a writer who interviewed me for an article on my beans misunderstood me and reported that I was growing a new bean called the "Rogue Mountain bean." Her article was apparently read by many people, and I occasionally see mention of a Rogue Mountain bean.

How Did We Get Thataway?

In August 1997 Berea College held an Agriculture Summit to discuss the future of its agriculture program. Speakers from several fields and from various institutions were invited to attend, and individuals from many departments on campus were in attendance as well. We spent the better part of two days listening to talks and holding discussions about possible alternative scenarios for Berea's agriculture program. I don't remember many of the speeches, but I did trade some heirloom tomato seeds with Wendell Berry, already a well-known Kentucky author who was one of the speakers.

What I do remember, vividly, was the banquet held on the last evening of the conference at Boone Tavern, the historic hotel and restaurant on the college campus. Probably close to a hundred people sat down to a typical banquet meal, which of course included green beans. The beans are the only thing I remember, and they made an indelible impression on me. The beans were straight as an arrow, very round, about five inches long, deep green in color, and, as I discovered, tougher than shoe leather. I couldn't puncture the beans with a fork or cut them, so, using my fingers, I picked up a whole one, put it in my mouth, and started chewing. An uncomfortable time later, I finally took it out of my mouth and, as discreetly as possible, placed the unchewable bean on the side of my plate.

Some time later I conducted a "scientific" experiment. After the banquet was over, but before the plates had been removed, I walked around the banquet room, checking all the tables and noting how many people had eaten the beans. The beans appeared to be untouched on every plate, while other portions of food had been eaten. If the banquet had happened earlier in the conference, we certainly would have had another interesting topic to discuss, one that says volumes about modern American agriculture and marketing and the gullibility of the American people.

Six years later I was giving a talk at a regional foods conference at the University of Mississippi. I noted the differences between "old-timey" beans and those produced through the miracles of modern plant breeding. I discussed how, in our efforts to reduce farm labor and maintain a "cheap food policy" popularized by major agricultural organizations, including land-grant universities, we were making some of our most basic foods almost inedible and of dubious nutritional value. I then told my story about the banquet beans.

Several chefs happened to be in attendance, and what they had to say was enlightening. To a person, they were well aware of the problems with commercial beans, but they served them anyway, just to adhere to tradition and to add color to the meal. Beans were used as a garnish, the chefs said; no one expected them to be eaten. That was a total surprise to me.

A much earlier incident in 1965 made a similar case against the modern tomato. I always planted at least ten times the number of tomato plants needed to feed my family, and I had a bumper crop. I went to Lexington to meet with the produce manager of Kentucky Foods, hoping to sell him some of my extras. The manager was quite pleased with the several tomatoes I had brought to show him. He sampled one of them, said it had excellent flavor, and offered to buy all the tomatoes I had to sell. So I went back to Berea, picked all my ripe and nearly ripe tomatoes, and took them to the assigned place the following day. I probably had close to ten bushels, and I was so proud of the tomatoes that I took a photograph of them before I left.

When I arrived back at Kentucky Foods and showed the produce manager what I had brought him, he became quite angry and belligerent and chastised me for my "ignorance." I had brought him ripe tomatoes, and he didn't want ripe tomatoes. He wanted green tomatoes that he could leave in storage until one of his stores needed tomatoes. He would then "gas" them as he hauled them to the individual stores, so they would appear red by the time they got there. He was absolutely correct when he called me ignorant. I had no idea that this was standard practice when it came to marketing tomatoes in grocery stores. I knew grocery-store tomatoes didn't taste very good, but it never occurred to me that I was eating green tomatoes that had been gassed to color them.

Of course, the tomato situation has become much worse since then. Modern plant breeding has produced tougher and tougher tomatoes that can withstand long-distance transportation, be kept in storage much longer, and still have a thirty-five-day shelf life when they finally reach our local megamarkets. It's no wonder that children boycott them in every possible way they can devise.

One other tomato story illustrates the depths to which we have sunk. My wife and I like fried okra with ripe tomato slices over the top. The tomato juice drips down into the okra, making it tastier. We normally don't use store-bought tomatoes, but she was going to fry some okra that had been frozen the previous summer, and I volunteered to go to the local megamart and get a large tomato to go with it. The tomato I brought home was very hard, so we decided to give it a few days to soften before we had our meal. Two weeks later, it was still as hard as a rock, but we decided to go ahead anyway. She cooked the okra and put it on our plates. I cut the tomato and put the slices on top. We sat down to eat, and before I took the first bite, I noticed that the tomato had a smell somewhat akin to formaldehyde, which led me to believe it might not taste so good either. It also had no juice to soak into the okra, which turned out to be a good thing. Neither of us could eat the tomato, and not wanting to be wasteful, I said I would feed the tomato slices to our son's chickens the next day. The chickens loved our homegrown tomatoes and always got more than their share. When I offered the tomato slices to the chickens, they ran over right away, but instead of immediately devouring the tomatoes within seconds, they took a whiff and ran in another direction. Two weeks later, the tomato slices were still lying on the ground where I had dropped them, seemingly indestructible. I assume they eventually decomposed.

Back to the Future

Those of us who subscribe to several agricultural and gardening publications have noticed that seed companies frequently change their names. This usually happens when a larger company buys a smaller one or when two or more companies join together to form a single larger company. Small companies have become a thing of the past, and many old-time plant varieties have become extinct. As seed companies consolidate and merge to form larger companies, the seed catalogs they send out to customers remain roughly the same size, but many older seed varieties that are not in demand are dropped. Numerous new tomato varieties appear, and they are often given numbers instead of names. If one numbered variety doesn't catch on with commercial growers or gardeners, it can easily be dropped and replaced with yet another numbered tomato.

Unlike tomatoes, new bean varieties are usually still given names, as there are much fewer of them. But it's hard to find descriptive adjectives for beans. They might be called tough, tougher, and toughest, since there is little difference in texture or flavor. When beans are described, a "sieve" number is sometimes given to indicate the size of the bean seed.

One name from the past can still be found in many seed catalogs: Kentucky Wonder. However, the modern Kentucky Wonder bears little resemblance to the bean of that name from decades ago, and the relatively new Bush Kentucky Wonder is one of the toughest beans I have ever run across. I tried it out the first year it was advertised and had an unfortunate experience. I had already sworn off commercial beans, but I had a relapse and decided to see what the new bean was like. I took the first ones I picked to the Berea Farmers' Market and sold some to an older Berea resident who loved beans, especially Kentucky Wonders. At the next market she came back and asked what had happened to her beloved bean. She said it was the

Growing for market: the Willard Wynn Yellow German tomato. (Dobree Adams)

toughest bean she had ever tried to eat and advised me not to sell any more at the market, since it would damage my reputation. I pulled up all the plants the following morning with the fervent hope that there had been no crossing with my heirloom beans.

The average person marching up and down the grocery aisles knows very little about the sources of the food on the shelves. And although today's shoppers might have had grandparents who gardened or who raised animals on a farm, most now live in suburbs and cities, where the closest farm animal is in a zoo and the closest live edible plant is in an arboretum.

The multinational seed companies are increasingly coming under fire for their breeding, marketing, and political practices and their attempts to control the world's food supply. There is also a growing movement to see that this doesn't happen. Many of Kentucky's small farmers, gardeners, and seed savers are doing their share to bring back the foods and gardening practices of the past that can help guarantee a healthier food supply in the future. Bigger is certainly not always better.

2

Kentucky's Seed-Saving Pioneers

All traditional seed savers have a love of the land, and for them, seeing things grow is highly satisfying. Sharing is important too, because once a variety becomes extinct, it can't be brought back.

I have known many seed savers who were far ahead of their time, but some stand out for their intensity and passion. Many have been lonely at times in their endeavors, but kept going nevertheless. They persevered, sometimes in the midst of ridicule and often in the face of skepticism from neighbors and even from friends and family. Here I introduce some of Kentucky's seed-saving pioneers. Some are still on the job; others I interviewed shortly before their deaths.

Vinson Watts

Vinson Watts, to whom this book is dedicated, was originally from Watts on Leatherwood Creek in Breathitt County. He graduated from Berea College in 1952, served in the Marine Corps, and then went back to Berea, where he worked as associate dean of labor under Wilson Evans. In 1956 Evans, who was born and raised in Lee County, Virginia, shared some tomato plants with Watts—a variety that had been in Evans's family for generations.

John Inabnitt of Pulaski County, a nephew of Nina Jones. A druggist in
Monticello, John is a big heirloom gardener with lots of beans, tomatoes, winter
squash, sweet potatoes, and other vegetables to his credit. He even has a heritage
breed of cattle.

Vinson Watts tomatoes often grow in handsome clusters.

Evans wanted to try some other varieties of tomatoes in his garden, and he asked Watts to take care of the Evans tomato and keep it pure.

Watts liked the tomato and decided to grow it exclusively in his garden. In 1967, when he became the first personnel director at Morehead State University, the Evans tomato went along with him. To keep the seeds as pure as possible, he grew only the Evans tomato in his own garden beside his house, but he grew other tomatoes in his neighbors' gardens. In addition, he continued to grow heirloom beans from his Breathitt County family.

Growing up during the Great Depression, when seed saving was a necessity, Watts knew how to save seeds from the Evans tomato. He chose only seeds from the most prolific, best tasting, and most disease resistant plants. Gradually, the tomato improved significantly.

Vinson Watts was a friend of mine when he worked at Berea

College, and I kept up with him after he moved to Morehead. I knew about his ongoing efforts with the Evans tomato, and he occasionally shared some of the seeds with me.

In 2006, at a seed savers meeting at my farm, I shared some of his seeds with another seed-saving friend, Merlyn Niedens from Illinois, who produced heirloom seeds for several specialty seed sellers. The next thing I knew, Vinson Watts tomato seeds were being offered by Baker Creek Heirloom Seeds, and Vinson found out about it before I did. He was flattered that his tomato was becoming so well known.

Shortly afterward, Andy Mead of the *Lexington Herald-Leader* wrote an article about Watts and his tomatoes, setting in motion a series of other articles in newspapers and magazines all over the country. People even came to visit Vinson's home. I had also started selling his tomato seeds through our website (www.heirlooms.org), and very quickly his seeds were outselling all my other tomato seeds combined. The Vinson Watts pink tomato usually weighs over a pound and has an outstanding balance of sugars and acids. A favorite of chefs at high-end restaurants, it has become my best-selling tomato at the Lexington and Berea Farmers' Markets.

In 2007 I left a row of Vinson Watts tomato plants growing in one of my high tunnels after the other varieties had already been removed. They grew to the top of eight-foot poles and back down to the ground, for a total of sixteen feet. The plants, set out on March 15, started bearing ripe tomatoes the first week in June and continued until I turned off the wood heat on November 22 and let the vines freeze. I was still picking sixteen-ounce tomatoes, but I had to clean up the high tunnels for the upcoming season.

Since I first started selling seeds of the Vinson Watts tomato, many other people have started growing the plants and selling the seeds, and the variety has become well known in the United States, Canada, and Europe. The tomato has become one of the favorite varieties discussed on Internet garden forums because it has a known history and is considered exceptional by so many people. For his work with his tomato, Watts was given the Lifetime Achievement Award by the Sustainable Mountain Agriculture Center in Berea in 2007.

During the last three or four years of his life, Vinson developed numerous health problems and was unable to do much with his tomato, other than grow it in his backyard garden and continue to save seeds from the best plants. Emphysema finally made him dependent on a portable oxygen tank. Even then, he would stand on his back porch and instruct the high school students who worked his garden for him.

Two days prior to his death, I asked Vinson when the Wilson Evans tomato became the Vinson Watts tomato. He was hooked up to tubes and unable to talk, but his mind was as sharp as it was the first day I met him. He kept a pencil and pad in his hands to communicate. To my question he wrote: 1980.

Vinson died on March 17, 2008. I potted one of his famous tomato plants, the sturdiest one I had, and his sons placed it alongside their father's casket, where it looked beautiful among the flowers and reminded those who had come to mourn his passing of the tomato plant that had consumed so much of his life.

I still have many of the original seeds he gave me and will continue to grow some of them each year, save the seeds from the best tomatoes, and share them with other growers. With any luck, I will be able to grow a few of his tomato plants from original seeds for many years.

Vinson's son Allen remembered his father with these words:

Few of us grow up along with a growing legacy. Now older, I am reminded each day of what it means to be the one who keeps something going for their own enjoyment and the enjoyment of people everywhere. Such a journey teaches us how to lead a great life. All this from heirloom tomato seeds? Yes indeed!

My father spent his entire life helping others, especially those who wanted to make for themselves a better life. The seasonal nurturing in his garden of his tomatoes is a mirror of his life's work. Constantly picking the best traits, year after year, to consolidate a most desirable product, he learned to work with nature, not against it. My father grew up gardening because he and his family had to for sustenance. When he no longer had to garden to have any food he did it because he enjoyed it, especially tomatoes. He passed this knowledge to his sons, grandsons, and friends who appreciate what it is.

Nina Jones

I first met Nina Jones many years ago when she entered some of her canned beans in a contest sponsored by Shakertown. The contest featured both fresh and preserved fruits and vegetables. Seeds also placed on exhibit were sold or given away. It is my understanding that Nina walked away each year with many first-place ribbons for her canned vegetables.

I interviewed Nina Jones two years before her death on March 29, 2012, just a month short of her ninety-eighth birthday. At the time, she was not very happy. The highway department had taken her house and garden in Somerset to make way for road construction. After demolishing her house and messing up her garden plot, the powers that be changed their minds and tried to sell the land back to her, after she had already bought another house a short distance away. Unfortunately, the new house lacked a good garden plot; there were only a few fertile strips where she could plant some beans and tomatoes and her ever-present flowers, but nothing like the large gardens she had always grown in the past.

Nina Jones of Somerset, Kentucky, at age ninety-six. She was a blue-ribbon winner for her canned beans.

Nina Inabnitt was born in Lincoln in Pulaski County, Kentucky, the second oldest of six children. Her family later moved to Stab, which was close to Acorn, both small communities in Pulaski County. After her older sister married and moved away at age sixteen and her mother became ill and unable to work, thirteen-year-old Nina pretty much took over running the household, tending the gardens, and looking after her four younger brothers. She regularly canned hundreds of quarts of beans, pears, and wild strawberries.

Nina didn't marry until she was forty-three years old, when she moved to Louisville and met Robert Jones of Louisiana. The two moved to Goshen, Indiana, where they both worked in the maintenance department of the local school system. Nina carried her bean seeds to Louisville and Goshen and continued to garden. She found good ground in Goshen and had excellent gardens there. After twenty-three years in Goshen, she and Robert moved back to Pulaski

County, bought a house, and immediately started gardening again. Her favorite bean varieties throughout her life were the Alice Whitis bean and the Agnes bean. These beans, along with several varieties of flowers, traveled with Nina to five different homes before she and her husband moved back to Kentucky and settled for good.

John Inabnitt

Now in his mid-fifties, John Inabnitt is a nephew of Nina Jones and lives a few miles from her former home in Pulaski County. A pharmacist by profession, John says that seed saving is both a hobby and a passion. He grows many heirloom beans and tomatoes and also raises British White Park cattle, a heritage breed grown for both milk and beef. A longtime member of the Seed Savers Exchange, he now confines most of his heirloom seed saving to Pulaski and Wayne Counties. He has more than fifty heirloom bean varieties from these two counties alone, in addition to several heirloom tomato varieties. He also maintains more than thirty heirloom sweet potato varieties, a pie pumpkin from the 1800s, and Walter Burdine dent corn, which is excellent for roasting ears. His seeds are available through a museum in Monticello across the street from his pharmacy and from him in person at his home.

For many years, Shakertown at Pleasant Hill held a festival in September that showcased heirloom seeds and vegetables and canned fruits and vegetables as well. John's collection was always the most outstanding. He and his aunt Nina, also a participant, dominated the festival.

Dorothy Phelps Inabnitt, John's mother, taught him to save seeds when he was a young child. He has continued the family tradition, trying to save as many seeds as possible in his corner of the world. John notes that by saving seeds from the most representative plants, a variety is being constantly improved, one of the most important benefits of seed saving.

Jim and Judy Tapley

Jim Tapley and his wife, Judy, live in Stearns, Kentucky, in McCreary County. One of many traditional gardeners in that county, he is best known for growing an heirloom potato, the White Sprout. He can trace it back at least eighty years to Lloyd Anderson. The White Sprout is a large, versatile white potato that is good for baking, frying, and making mashed potatoes. At a time when most gardeners buy their seed potatoes from large suppliers and grow mostly Red Pontiacs, Irish Cobblers, and Kennebecs, Tapley has doggedly hung on to an older variety that, in many respects, is superior to all the commercial types available.

Jim Tapley of Stearns, Kentucky, is best known for growing the White Sprout heirloom potato.

Judy Tapley of Stearns, Kentucky, cans several hundred quarts of heirloom beans each year and shares them with friends, neighbors, and relatives.

Jim and Judy also grow and can several varieties of heirloom beans, especially the Herman Hatfield Half-Runner, which they started growing in the 1970s. Their house is near the top of a hill, and their gardens are planted down the slope on terraces, which keeps the soil from washing away and maintains its fertility. Jim's potatoes are stored very effectively in a root cellar on the hillside, and they look just as good and firm in January as they did when harvested in July. The Tapleys eat the larger ones and keep the smaller ones for seed, planting them whole instead of cutting them up. Each summer, Judy cans several hundred quarts of heirloom beans, sharing them with friends, neighbors, and relatives in northern Kentucky.

Like many people from the McCreary County area, Jim and Judy went to Chicago for two years and worked in the factories, only to come back to McCreary. Jim then worked in the coal mines. Now in his late sixties, he is disabled with emphysema but still able to raise his gardens and maintain his heirloom seed stock. Judy works for the McCreary County school system, helping the youngest students with learning disabilities.

Zeke Dishman

Zeke Dishman was raised in the Windy area of Wayne County and still lives there. A lifelong logger, he has also gardened intensively all his life. Several decades ago, while Zeke was working on a logging operation, a woman who lived nearby gave him a tomato variety she had been raising for many years. He took it home and continues to grow that tomato to this day.

This variety, known locally as the Zeke Dishman tomato because of his long history of growing and improving it, is one of the best red heirloom tomatoes available. It bears large fruit that often exceeds two pounds and has an excellent flavor and very few seeds. Having the acidic flavor of most red tomatoes, it is good for eating fresh, for cooking, and for canning. And because it has few seeds for its size, it is a very good tomato for individuals who must avoid small seeds in their diets for medical reasons. Even though he has been disabled by a stroke, Dishman still maintains his garden and grows the tomato he has worked with for so long.

J. B. Mullins

J. B. Mullins lives in the southwestern part of Breathitt County, where he farms over 500 acres on Turkey Creek. Like so many other traditional gardeners in Kentucky, Mullins migrated north but didn't stay very long. After serving with the 8th Army Engineers during the Korean War, he worked for a Michigan steel mill from 1954 to 1966 before heading back home to farm.

Mullins was an orphan, and his adoptive family saved seed by necessity. His love of seeing things grow developed into a particular love of beans, and he has accumulated more than fifty heirloom varieties. He is also fascinated by apples and maintains more than fifty "heritage" varieties. He grafts as many of the old apple varieties as he can find, including Winesap, Arkansas Black, McIntosh, Yellow Transparent, Northern Spy, Red Delicious, Wolf River, Horse, and Spice.

Some of Mullins's beans are well known in the Breathitt County area, where he sells seeds at the Ace Hardware store in Jackson. Because of a very poor growing season in 2009, he didn't sell any seed beans there in 2010. He had lost many beans to heavy rains and had managed to save only enough seeds to keep his own stock going.

Now over eighty years old, Mullins doesn't want to lose any of the varieties he has maintained for so many years. Therefore, he has decided he must plant many varieties together in a single row. He typically grows about fifteen bean varieties per year. Some of his more interesting beans include the Daddy bean, which was brought to the area by river raft, named after the man known as Daddy to his river crew. The Jane Herald bean is somewhat like the original Missouri Wonder. The Dog Eye bean looks like a dog with eyes of two different colors, and the Turkey bean was originally taken from the craw of a wild turkey. Mullins was somewhat surprised to learn that many gardeners like to grow his mixed beans because they like the flavor of so many different beans cooked together.

J. B., who has pastored a local Baptist church for many years, starts many of his seeds in a greenhouse made from recycled glass. He also grows heirloom tomatoes and starts beans and tomatoes in flats in his greenhouse to get an early start on the growing season.

Joe Richards

Joe Richards, originally from the Ozark community in Adair County, has lived in Pulaski County for decades. He worked for the Pulaski County school system for thirty years, teaching business and accounting courses for six years and then transferring to the central office. His wife, Betty, taught math and science for

Joe Richards of Somerset holding seeds of the Conover butter bean.

thirty years and then served on the school board for twenty years. Joe received his college education at Lindsey Wilson College and later did graduate work at Union College and Eastern Kentucky University.

Throughout all his years in the Somerset area, Joe has been an avid gardener, maintaining family heirloom vegetables and collecting others, as well as being a member of the Seed Savers Exchange. In addition to beans, his collection includes the Whippoorwill field pea, a giant sunflower, a plum granny, and the ornamental plant Job's tears, which many people use to make necklaces. Now in his mid-seventies, Joe has been retired for some time, and health problems have caused him to slow down, but he still has some of the best land for gardening in Pulaski County. Joe talks about going into Russell County to cut wild cane growing along creeks. He uses the cane as poles to grow his butter beans. He carefully plants a few beans around each cane, taking care to ensure that all the different beans are represented in each planting. When the growing season is over, the canes are stored in a dry place and used again the next season.

Joe Richards comes from a long line of gardeners that can be traced back to the Civil War. His great-grandfather, William Henry Conover, was a Union soldier whose unit was stationed near New Orleans when the war ended. The troops had to walk back to Kentucky to resume their lives—a trip that took a long time. Conover decided to bring some beans home with him, and whenever he walked by someone's garden, he picked a few butter beans. By the time he made it back to Adair County, he had quite a collection of butter beans in a variety of colors. His descendants have been growing them ever since. Few heirloom beans can be traced back as far and as directly as Joe Richards's butter beans. As far as Joe knows, no attempt has been made to separate the butter beans into different varieties. He believes he is the only descendant of his great-grandfather still growing the beans.

Cliffie Strong

Cliffie Strong, a ninety-five-year-old widow living in Owsley County at the time I interviewed her, had a remarkable collection of beans and other seeds from her gardening career, which lasted from 1934 through 2006, when poor health kept her from working in her garden. However, at the time of my visit she was still hoping to regain enough strength to start again. Cliffie died in early March 2013 at age ninety-seven.

In addition to growing gardens for her family, Cliffie hired out to do farmwork for other people when time permitted, earning fifty cents a day. Still, she did a massive amount of work on the farm belonging to her and her husband, as well as canning and drying most of the food needed by their family. This included stringing, breaking, and drying the twenty or more bushels of beans needed to make five bushels of shuck beans (due to the shrinkage in volume as the beans dried).

Bean seeds still in Cliffie's freezer after her death included Little

Cliffie Strong, matriarch of four generations of seed savers.

White Bunch, Big White Bunch, Big Brown, Brown Goose, Gray Goose, White Goose, Striped Goose, White Case Knife, Tobacco Worm, Pink Tip, Crane Toe, Golden Hull, Six Week Cream Colored, Fat (a fall bean), Nannie Coulton, and Dry Weather. The Nannie Coulton is a mutant bean found by Cliffie's friend Nannie Coulton in her garden. Cliffie was so satisfied with her own beans that she never even tried half-runners when they gained considerable popularity some fifty years ago. In addition to her many varieties of beans, Mrs. Strong grew black-eyed peas, Whippoorwill peas, and the Sugar Crowder pea. She also grew "Old Flat" mustard, white sweet potatoes, and Irish potatoes.

Donna Morgan, Cliffie's granddaughter-in-law, wrote the following a couple of years before Cliffie's death:

Upon meeting Cliffie Boyd Strong, you can expect a bright smile and a handshake, often accompanied with her special introduction: "Hello, I'm Cliffie Strong—very strong." Cliffie Boyd Strong resides in Owsley County, Kentucky, on Big Springs Road. She was born in that area on October 5, 1915. Along with her daughter, Lucille Lamb, she maintains the farm where she and her husband, Vernon, raised their four children. Cliffie tells the story of how when she and Vernon first married, they had a rather lean year. After that first year of marriage, she said she decided they would never be hungry again, and she began to raise a substantial garden. Any meal at Cliffie's home is sure to include either canned or shuck beans, corn, tomatoes, potatoes, and other vegetables raised right up the hill from the house.

Since her first garden, Cliffie has saved her heirloom seeds for Kentucky "Case Knife" beans, cut-short greasy beans, old-fashioned mustard, and other vegetables. She comments that she saves only the best full-seed pods to ensure quality seeds year after year. In addition to being a master gardener, Cliffie is an accomplished wood carver and quilter.

While Cliffie is able to play a minimal role in the gardening at this point in her life, her daughter Lucille continues to raise the beans, mustard, corn, and other plants on the family homeplace. Lucille returned home to be with her mother following a career in the army. Meanwhile, Lucille's son, James "Moose" Morgan, and his family raise the heirloom beans and mustard as well. The Morgans reside in Estill County, Kentucky. The Morgans' children, Sean and Amy, also participate in gardening. As Cliffie's great-grandchildren grow up and move on to their own lives, they expect to carry on the family's traditional seeds.

Frank Barnett

One of the youngest and most active of the old-time seed savers is Frank Barnett of Georgetown. Now in his late sixties, Frank is on a mission to visit gardeners and vegetable stands in eastern Kentucky to make sure he has all the beans being grown or sold in a particular area.

When health problems in the late 1990s caused Frank to reconsider his diet, he stopped eating out of cans and began growing a lot more of his own food. His first visit to the Lexington Farmers' Market in 2000 reminded him of his grandma Barnett's mottled bean, which she had grown for most of her life. The heirloom beans he bought at the farmers' market inspired him to add to Grandma Barnett's bean. After he retired from IBM in 2007, Frank began devoting a major portion of his time to collecting, growing, and sharing heirloom beans.

Frank has narrowed his search to southeastern and south-central Kentucky and adjoining counties in Tennessee, Virginia, and

Frank Barnett at the annual Pikeville seed swap, 2015. (Dobree Adams)

West Virginia, primarily in the coalfields. He found that family heirloom beans were few and far between in the parts of Kentucky where
large farms are predominant, but within a hundred-mile radius of his
boyhood home in Floyd County, it was hard to find extended families that didn't have a family bean. He has also been successful in
locating family heirloom tomatoes.

As of 2014, Frank had more than 400 heirloom bean varieties, over half of them from Kentucky. Regrettably, he has only one of his grandma Barnett's heirloom beans, although he knows there were more. To maintain the purity of his beans, he spreads his plantings over several patches of land at his home and on two farms.

Here is Frank's story in his own words:

I started collecting heirloom beans with my grandmother's cornfield bean after she passed away in 1990. I had also raised commercial White Half-Runners in the past, but after visiting the Lexington Farmers' Market and discovering the beans from Bill Best and Dwight Evans, I threw all my commercial seeds in the garbage.

Searching for heirloom beans in Kentucky was very discouraging until I started looking in southeastern Kentucky after my retirement. Gardeners know their beans and know how to raise them by using wooden sticks, cane poles, or trellising.

In Menifee County at the Freewill Pentecostal Church at Means, I met Larry Sexton and Marvin Cassidy, who were raising a two-acre garden on the church grounds. They had a striped greasy bean that Larry showed me how to pick. Up to ten pods would form on a limb in a "soldier" formation, which made it easy to grab a handful at a time. I bought a half bushel, which gave me plenty to eat and more than enough to save for seed. It is an excellent bean.

In Wolfe County I stopped at the True Value Hardware store in Campton owned by Darrell Halsey. He had the standard commercial seed for sale, but he was also selling seeds of four local beans that he raised. I bought a good portion of each bean, and Darrell gave me the names of the people who had originally given him the seed.

In Breathitt County at Jackson in 2008 I had stopped at the produce market downtown on R15 operated by Jerry Fraley. To my surprise, he was selling beans that had been strung and broken up. I bought an excellent speckled greasy cut-short bean from him, which I grew in 2009. Jerry was selling at his garden site on R30, three miles west of Jackson, in 2009, having sold his market in Jackson. I had to ask him how many beans he and his friends had broken and strung the prior summer. He told me fifty to fifty-five bushels!

I stopped at the downtown Jackson produce market again, and the new owner was buying beans from a local grower. He had three bushels of beans—a striped pole, a speckled pole, and a creamy bean that was a family heirloom. The new owner didn't want to buy the creamy bean, saying it was easier to sell the "big long pretty" beans. Of course, I bought a good portion of each of the pole beans and the creamy bean after the grower took them back to his truck. I was able to save plenty of seed of each bean.

There is a produce stand at the top of the hill beyond Walmart on R15 on the way to Hazard operated by Rebb Hudson. He buys from local growers for resale. I have bought several good beans from him.

At the True Value Hardware in Jackson, I was able to buy some local heirloom bean seed grown by J. B. Mullins of Turkey Creek. Last summer I grew his Jane Herald Pole and World War II Victory beans. However, due to the floods of 2009, J. B. Mullins could not supply beans to the hardware store the next summer.

In Perry County in 2008 I stopped at Smitty's Market on Combs Road in Hazard. They go twice a week to the Western North Carolina Farmers' Market in Asheville to buy produce for their market. They also grow produce in their own local gardens to sell at the market. In fact, they close down the market the last three weeks of August just to bring in their own gardens.

I was able to buy mature pods of a white cornfield and a long speckled greasy bean they had bought at Asheville. They also gave me a seed packet of the Maggie bean, which they raise to sell.

In 2009 I started going to the Hazard Farmers' Market. I met Oma Clark of Owens Branch, who was selling her own greasy bean. After I told her I was looking for old-time beans to save for seed, she gave me all the yellow mature bean pods she had left. She wouldn't accept any payment but mentioned that she was looking for an old bean she had lost years ago, which sounded like my grandmother's bean, so I got her address and mailed her a bag of my grandmother's beans the following Monday.

James and Sarah Birchfield of Wolfe County were at the market selling white greasy grit beans for $8 a gallon or $60 a bushel. I bought a gallon. They were good beans, but they really earn their money picking the beans, since they are short and small. However, they didn't have any trouble selling the bushels they had in their pickup truck.

In Floyd County the place for beans is the Bull Creek Trade Center (flea market) on R23 between Prestonsburg and Allen. They sell local beans if available, but most of the vendors buy their beans out of east Tennessee or western North Carolina.

In Morgan County there are several markets, but they sell out quickly. In 2009 I lucked out at the Morgan Sorghum Festival at West Liberty. I was able to buy seed from a vendor who had beans from Wanda Hamilton of Coffee

Creek. These included a fall bean, a striped cornfield, and the Square House bean (looks like a White Half-Runner type).

There is a real opportunity to collect heirloom beans in southeastern Kentucky. Many gardens were flooded out in 2009, but several people told me that in past years, anyone with beans would be selling on R15 all the way to Whitesburg.

Frank Barnett's description of his seed-saving activities reveals many things about Appalachian culture—selling vegetables along the side of the road, closing down one's store to harvest one's own crop, trading seeds for store goods and supplies, swapping seeds, traveling to another state to buy beans that are unavailable locally, and the overall sense of connectedness of all the participants. Even though he spent his career working for a major international corporation, Frank maintained his cultural roots, and in retirement he is providing a great service by helping to maintain the foods of his youth, which will be increasingly important in the future. He is also helping people realize the value of what they have.

3

Getting to Know Beans

Suppose that a few hundred years ago, a tribe of Native Americans sent a smoke signal to a nearby tribe inviting them over for a supper of "grass-like" beans that had been picked before "lumps" appeared in them. Or suppose that the message assured the guests that the beans had been picked while they were "young and tender" and were "guaranteed not to stick to your ribs." These messages seem absurd to those who know beans about beans, but that is essentially what most seed companies tell would-be buyers of bean seeds every spring. The humble bean, so long an integral part of the diets of Native Americans and those who later joined them in the New World, is under attack by those who have genetically modified beans to make them tough, tasteless, protein free, and easy to harvest by machine. We seem to have forgotten that beans were once a primary source of protein in the diet. And the source of that protein is the seed, the part of the bean forbidden to develop in the modern bean prior to being harvested for food.

Prior to the advent of machine-harvested "bush" beans, the vocabularies of bean growers, which included almost everyone, were full of descriptive names such as cornfield beans, fall or October beans, cut-short beans, wax beans, string beans, greasy beans, pink-tip beans, dry beans, shelly beans, butter beans, shuck beans, fodder

The beautifully speckled Josara Fall bean. (Frank Barnett)

beans, and a host of others. For most older Americans, the bean of memory is the cornfield bean. Corn, grown for human consumption and for livestock, also provided support for cornfield beans. The beans simply climbed up the cornstalks and, with their relatively broad leaves, captured enough sun to produce a large crop of beans. Open-pollinated varieties of corn had strong stalks and were tall enough to support bean vines that often grew to over sixteen feet.

Beans had a symbiotic relationship with corn. While the cornstalks provided support for the beans to climb, the beans, being legumes, fixed nitrogen in the soil for use by the corn plants. Historically, pumpkins and winter squashes were also grown among the cornstalks, helping to keep the rows somewhat free of weeds and providing additional food for humans and livestock. The Native Americans called corn, beans, and squash the "Three Sisters," and these plants provided a good diet, even in the absence of meat from game animals. Meals consisting of beans, corn, and squash would "stick to your ribs," as the old mountain saying goes.

But as humans became "wiser," hybrid corns were developed to increase the yield of grain to feed animals or, more recently, to make ethanol to fuel cars. The focus was on short, early-maturing stalks that weren't strong enough or tall enough to support vining beans and barely strong enough to support an ear of corn. Fodder, the portion of the stalk above the top ear, was no longer being used to feed livestock, so a tall stalk was no longer considered important.

Many gardeners ceased to use open-pollinated corn to make meal or hominy, and some stopped growing corn in their gardens altogether. If they grew corn, it was the modern, "sweet" hybrid varieties with short, weak stalks. This left beans without any support, and cornfield beans, of necessity, became pole beans (or, as some people call them, stick beans). Those who lived near canebrakes could cut the long cane poles and use them to support climbing beans. The canes were often as tall as the cornstalks had been, and when arranged in a pyramid or "teepee" style, the poles could support a lot of weight and resist wind to some degree. Some enterprising gardeners cut tree limbs with lots of small branches and buried the trunks in the ground, allowing the bean vines still more freedom and space to grow.

More recently, vining beans are grown on trellises created by posts, wires, and strings to provide an optimum amount of space and sunlight. Posts are sunk into the ground about eighteen inches deep every twenty feet or so. High-tensile wire is stretched overhead and secured at the ends to keep the posts from leaning under the weight of the vines. String is then tightly secured to the bottoms of the posts and twined up and down in a zigzag fashion to create a trellis. Cornfield beans, which became pole beans, are now often trellis beans, although they are still grown on open-pollinated corn and on poles as well.

For the really serious gardener, page wire fencing is sometimes secured by steel T-posts. Concrete reinforcing wire is also used, stacked at double height. And for real strength, some gardeners use cattle panels stacked two high and secured with steel posts or strong wooden posts. When all is said and done, however, cornfield beans are still cornfield beans, regardless of how they are supported during the growing period. Interestingly, many people who speak of cornfield beans have no idea how that term came to be.

Commercial seed companies typically sell two types of beans: bush or bunch beans and pole beans. Bush beans usually outnumber pole beans by about ten to one, and some catalogs offer no climbing beans at all. These catalogs picture bush beans as being straight as an arrow, as if that had anything to do with anything (actually, it is important for mechanical harvesting). And most companies confine themselves to no more than thirty varieties, sometimes far fewer.

In contrast, heirloom beans, and certainly those of the southern Appalachians, are mostly cornfield beans, with only a tiny fraction being bush beans. A hundred years ago, information about cornfield beans was common knowledge, but today, probably not one in ten thousand people possesses the knowledge that was almost universal just a few decades ago. When large companies took over seed saving and distribution for the rest of us, much of our cultural memory and wisdom passed away, and as certain types of beans were declared obsolete, that part of our culture became obsolete as well.

Just Plain Cornfield Beans?

When most people speak of cornfield beans, they mean climbing beans. These beans vary greatly in type, and they come in many colors and shapes. Most heirloom cornfield beans also have strings that must be removed prior to preparation for eating—one of those things people used to know without being told.

Several years ago, a young lady visited my stand at the Lexington Farmers' Market and asked to buy some cornfield beans so that she could prepare a bean dish her southern grandmother used to make. She was a flight attendant for one of the airlines served by the Lexington airport, so she bought a few pounds of beans and took them home on the airplane with her. A week or so later, she came back to the farmers' market and asked me what she had done wrong: her beans had been filled with strings. When I explained that the strings had to be removed prior to cooking, her memory was jogged a little, and she remembered that part of the process from her childhood. She practiced stringing a bean in my presence and then bought some more to try again.

Hundreds of other customers at the Lexington Farmers' Market regularly buy cornfield beans for cooking in the traditional ways of their ancestors. Many buy them for canning, and a significant number buy them for making shuck beans (also called shucky beans, leather britches, and, especially in Clinton and Wayne Counties, fodder beans). Preparing shuck beans has become something of a lost art, but it is still important in many families. Shuck beans are simply dry beans with the hull intact. They are usually made from "full" beans—that is, beans in which the seed is nearly fully mature. Historically, they were a primary source of protein, and a bean that consisted mainly of hulls wouldn't make the grade. The strings are removed, and the beans are left whole or broken into pieces, depending on the manner in which they will be dried. If the beans will be strung with a needle and thread and hung up to dry on a porch or in a room in the house, they are usually left whole because of the time it takes to string them. Shuck beans that are strung and then hung up generally dry quite readily. If the beans will be dried outside on a window screen or bedsheet, they are typically broken into pieces, just as one would prepare green beans. They are usually brought indoors at night to prevent the dew from falling on them. Placing the beans on a hot tin roof during sunny days speeds up the drying process. For those with access to a sunroom or a greenhouse, the beans generally dry in about a week with several days of sunny weather, and they typically retain much of their original color.

Cut-Short Beans

Cut-short beans are among the most popular beans grown by traditional gardeners in Kentucky. As the name implies, these beans have been cut short. As the pods grow to maturity, the seeds tend to outgrow the pods, making them very tight, and the seeds become flattened on the ends. The flattening process is so pronounced in many varieties that the seeds take the shapes of triangles, squares, rectangles, parallelograms, and trapezoids.

When cut-short beans are cooked, the seeds often break free from the hulls, resulting in a pot of mixed seeds and hulls. This is just what cut-short lovers want. All the cut-short beans I have seen are tender to an extreme degree.

When saving cut-short beans for seed, it is important to time the harvest carefully. Once the hulls have become dry, the bean pods need to be picked as soon as possible. If the dried beans are subjected to rain or even heavy dew for several mornings in a row, the pods often break apart and scatter the seeds on the ground—hence the term "bust-out" beans for cut-short beans.

Greasy Beans

Grown throughout the southern Appalachians and especially in eastern Kentucky and western North Carolina, greasy beans occupy a special niche among heirloom beans. They are considered the best of the best among bean lovers because of their tenderness and flavor.

The hulls of greasy beans lack the tight-knit fuzz of other beans, so they tend to shine and appear greasy to the eye. They range in size from cut-short greasies such as the three-inch-long Ora's Speckled bean from Jackson County, Kentucky, to longer greasy beans like the Striped Hull Greasy Cut-Short and the Mary Moore Greasy, also from Jackson County.

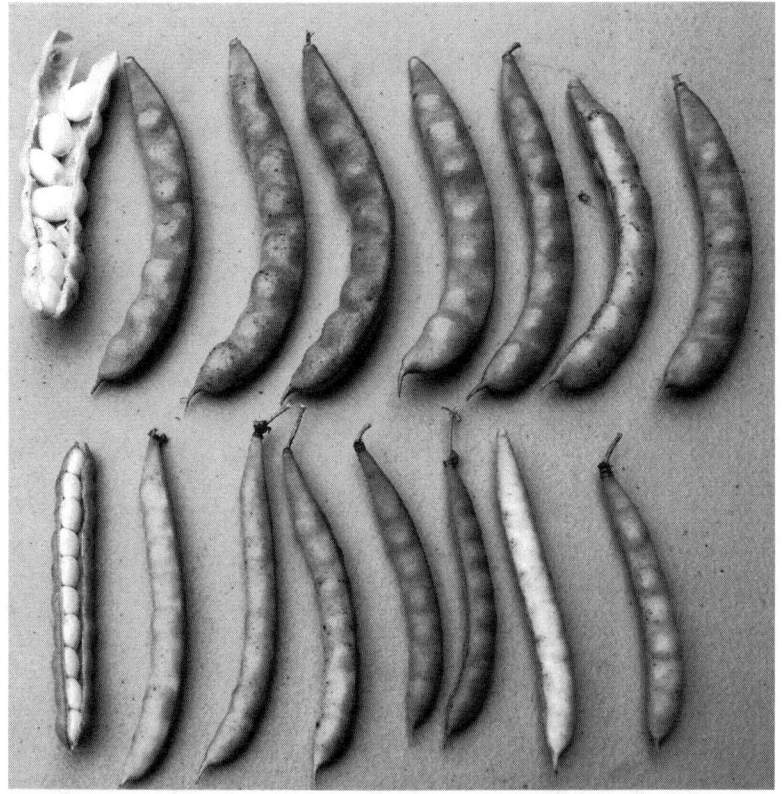

Two varieties of greasy beans from my extended family, showing what a "full bean" looks like.

Greasy beans come in many colors, with white being the most common. Some are brown, many are speckled or mottled, and a few are black. Mottled beans are typically a light brown color with darker brown stripes, specks, and smudges without any predictable pattern. A very few mottled beans have black stripes and smudges. Ora's Speckled bean has a light brown seed with darker brown specks. The Striped Hull Greasy Cut-Short and the Mary Moore Greasy are both white-seeded beans.

Greasy beans are very popular at farmers' markets, where they

Striped Hull Greasy Cut-Shorts from Jackson County, one of my best beans over the years.

typically sell at five to ten times the price per pound of commercial beans and twice that of other heirloom beans. The downside to greasy beans, if there is one, is that they take much longer to harvest than larger beans.

Greasy Cut-Shorts

To many people, heirloom beans that are both greasy and cut-short are the best of all—exceptional tenderness, flavor, and texture and a superior protein content because of the high proportion of seed to hull. Roughly half of all greasy beans are also cut-shorts, but the same is not true of cut-shorts. With a piece or two of cornbread, butter, onions, and fresh tomato slices, greasy cut-shorts make a meal fit for kings and queens. Some people believe the term "greasy grit" comes from small greasy cut-shorts that look like grits when dry because none of the beans are rounded.

Pink Tips

There are many varieties of pink-tip beans, but they all have one characteristic in common: the tips—the part of the bean most distant from the vine—turns pink as the bean nears maturity. Most people who grow pink-tip beans won't harvest them until the tip changes color.

The seeds of pink-tip beans can be just about any color from white to black and everything in between. I know of only one variety of greasy bean that has a pink tip, but many other cornfield beans have this characteristic. Sometimes the tip turning pink is a prelude to the entire hull becoming pinkish, but most of the time the tip is the only part of the bean to change color. Very occasionally a blue-tip variety shows up, but I have only one blue-tip bean in my collection.

Fall Beans

Fall beans, sometimes called October beans, are typically grown late in the season and are used primarily as shelly beans or dry beans. A shelly (or shellie) bean is cooked just after it has been hulled out from a bean that is fully mature but has not started to dry. Many people freeze them raw and undried in plastic bags or in jars and use them as soup beans. Others wait until they are dry; store them in a cool, dry place or in the refrigerator or freezer; and then rehydrate them and cook them like other dry beans, such as pintos. If the beans are not eaten by the time warm weather arrives in the spring, it is important to refrigerate or freeze them to prevent infestation by bean weevils.

There are a few tender-hulled fall beans that can be eaten without shelling them first, and many people can these varieties as well, still enclosed in their hulls. And some beans, such as the Paterge (Partridge) Head, are excellent prepared in any of the above ways.

Fall beans range in color from solid white to deep red. Many are speckled, some are mottled, and some are white with red, black,

Jim Gilbert of Harlan County showing his method of tying strings to overhead wires for his tall fall or October beans.

brown, or yellow eyes. Others are multicolored. Some are half white and half black with unpredictable patterns, such as the Baby Face Fall bean. Some are full runners, others are half-runners, and some have very short runners. At one time, most traditional gardeners raised at least one fall bean.

There are some bush varieties of heirloom fall beans, but most are climbing beans. The bush varieties that I know about have yellow coverings on the pods with reddish markings, similar to what are called "horticulture" beans in some parts of the country.

Butter Beans

Like fall beans, a lot of gardeners once grew at least one variety of butter bean. The hulls of butter beans cannot be eaten, but the beans can be shelled out and eaten as shelly beans before they are dried. Or they can be dried to be eaten later in the winter.

Butter beans vary greatly in size and color. Many are white, but they can be every color of the rainbow. People often grow them in a mixed form with as many as ten or more colors. Butter beans typically require longer to mature, and it is a good idea to allow 120 days from seeding to harvesting dried beans.

Wax Beans

Wax beans have a yellow, waxy hull and are not nearly as popular as other types. Many people use them in three-bean salads, but wax beans are not held in the same high regard as other heirloom beans in Kentucky. I don't know of many people who grow them, but those who do often like them very much.

To Bean or Not to Bean; That Is the Question (With Apologies to William Shakespeare)

As a lifelong grower and eater of beans, I have paid a lot of attention to them over the past fifty years while growing beans for the fresh market and over the past twenty-five years while also growing them for preservation and seed sharing. I never cease to be amazed by beans. It is always exciting to find a new bean in your patch, a bean that didn't exist before. It is exciting to try it out the following year to see if it comes back true and, if it does, to give it a name and share it with other bean lovers.

To bring a bean back from the brink of extinction is exciting, as I have done over the past few years with a bean taken from West Virginia to Oregon in the 1890s. The bean had been grown for decades by members of the Noble family until a lapse about twenty years ago, when, for some reason, no one in the family was still growing it.

What about Modern Beans with Designer Genes?

As far as I can tell, only two seed companies have dropped me from their mailing lists, so I still get more than my share of colorful seed catalogs. I dutifully read about new developments in bean breeding to see what "improvements" have been made.

In recent years, the new beans being advertised have a deeper green color, which seems to be what customers are demanding, or at least that's what gardening magazines tell us. They are still straight as an arrow and held high off the ground near the tops of the plants to facilitate machine harvest. We are still supposed to pick them when they are young and tender, before any lumps (seeds) appear. The company that once advertised its beans as being "grass like" no longer does so, since that probably wasn't the best image to use. (After all, why grow beans? Just mow your lawn and put the clippings in a pot.) It would probably not be a good idea to advertise them as being "protein free" either.

In the meantime, a quiet revolution is taking place as many thousands of people discover and rediscover heirloom beans. Although most of these beans require staking or trellising, and most of them must be strung prior to cooking, the end product is worthy of the name *food*. It is a journey into the past that is worth taking and one being taken by more and more people.

4

Getting to Know Tomatoes

There are not nearly as many tomato varieties in Kentucky as there are bean varieties, but we have some of the best, especially our heirlooms.

We used to be able to look in commercial seed catalogs and find a multitude of tomatoes for many purposes. The seed companies catered to gardeners by offering a host of tomatoes of different sizes, colors, and shapes. Some of the early hybrids were bred for flavor, texture, and nutrition. And many tomato varieties were good for canning as well as eating fresh.

However, about sixty years ago, retail grocery managers discovered that tomatoes could be picked green, kept in storage for some weeks, and then gassed en route to their destination to give them color. At the time, these tomatoes still retained some flavor, even after acquiring color through the use of ethylene gas. But things quickly began to change.

As seed companies merged with and acquired one another, many of the early tomato varieties fell victim to the bottom-line syndrome: if they weren't profitable, they were dumped. Plant breeders were instructed to produce tomatoes that could be shipped farther and farther from the farm, stay in storage longer, and still have a long shelf life. Plants were reduced in height and

The Zeke Dishman, the largest red tomato I grow. (Dobree Adams)

designed so that all the fruit would ripen at the same time. Indeterminate tomato vines gave way to determinate ones with high yields.

At the same time, flavor, texture, and nutrition took a backseat to durability, and home gardeners found themselves taking a backseat to commercial growers with lots of acres, migrant workers, and political clout. Grocery chain stores and fast-food restaurants came to dominate the market for commercial tomatoes, and the seed companies fought to humor them. Seed companies became larger and fewer, and they acquired international customers as well.

The first genetically modified food was the Flavr Savr tomato in 1994, which was said to maintain its flavor for a period of weeks. There was only one problem: in order to save flavor, there had to be flavor to begin with. The new tomato, bought by Monsanto, lasted only three years. Millions of dollars had been wasted.

Heirlooms Make a Comeback

About thirty years ago, many people had become fed up with the sorry state of modern tomatoes and started hunting for older varieties, especially as the best of the hybrids, such as the Ramapo, were discontinued. Growers who sold at farmers' markets actively sought these older varieties from traditional gardeners and started to grow and market them. Heirlooms, as they were called, moved slowly at first, but then the movement picked up and became an avalanche. Kentucky and other nearby states led the charge.

Some of the older open-pollinated commercial varieties, such as Campbell's 1327 and Rutgers, were also called heirlooms. Both have excellent flavor and texture and were originally bred for use by home gardeners and canneries. But a significant number of gardeners were looking for even older varieties that had been developed by gardeners and kept alive for their flavor and texture alone. They found them in abundance in Kentucky.

Types of Heirloom Tomatoes

Just like beans, tomatoes come in many types, sizes, colors, flavors, and shapes. The four basic types of tomato are table, canning, paste, and tommy toes. Table tomatoes typically weigh from six ounces to three pounds. Canning tomatoes are smaller and have a more intense flavor. Paste tomatoes, which have few seeds

Vinson Watts tomatoes on my favorite rock on our porch.

and thick walls, are generally used to make ketchup and sauces. Tommy toes are the old-fashioned small tomatoes with an intense tomato flavor. Modern hybrid cherry tomatoes, which have been bred to be sweet and to have a fruit-like flavor, have pretty much lost the acidic flavor of traditional tommy toes.

Heirloom tomatoes vary greatly in size, from tiny tommy toes, several of which can fit into a teaspoon, to large beefsteak types that can weigh up to seven pounds. Slices of some larger tomatoes can cover an average-sized plate and dwarf a slice of bread.

As the saying goes, there are more colors of tomatoes than you can shake a stick at. They range from solid white to solid black with dozens of shades in between. Red tomatoes are still the most popular, since there are more varieties to choose from. Other popular colors are gold, yellow, bicolored (typically yellow tomatoes with red stripes), pink, tan, green-when-ripe, purple, and black.

As a general rule, color seems to determine the flavor of a tomato or at least be indicative of it. The whiter tomatoes tend to have more sugars than acids, while the darker ones tend to have varying ratios of sugars and acids. Red tomatoes tend to be the most acidic, while pink tomatoes are high in both sugars and acids—often referred to as "old-fashioned" flavor. Yellow German tomatoes (yellow and red bicolored) tend to be the sweetest of all large tomatoes and are prized by many people for that reason. White, tan, gold, black, and green-when-ripe tomatoes all tend to have complex combinations of sugars and acids, and many people have a decided preference for certain colors, especially the black ones.

Most people think of a tomato as being round, but heirloom tomatoes come in many different shapes. Many are round, but there are also pear, tube, and heart shapes. Some are shaped like peppers and are even hollow like peppers, with the seeds clumped together in the center. These make very good stuffing tomatoes.

Heart-shaped tomatoes, called oxhearts, tend to be very flavorful and meaty and have few seeds, making them especially good for people who cannot tolerate tomato seeds in their diets. They come in many colors, but the pink ones seem to be the most popular at farmers' markets. Tube-shaped tomatoes, like the commercial Roma types, are usually used for processing into sauces. But unlike commercial Roma tomatoes, heirloom ones are as tasty as table tomatoes, especially the larger ones, which are used for sandwiches and salads as well.

Many tommy toe tomatoes are pear-shaped, with necks of different thicknesses. There are yellow pear, red pear, and black pear tommy toes, as well as round ones that are red, yellow, gold, white, cream-colored, or green.

As one might expect, heirloom tomatoes come in hundreds, if not thousands, of varieties. There are certainly hundreds of vari-

An assortment of shapes and sizes at the Lexington Farmers' Market in the summer of 2015, including Aunt Cecil's Green, Vinson Watts, Basin Mountain Tommy Toe, and Russian 117 Bicolor tomatoes.

eties in the southern Appalachians, and Kentucky has more than its share. Many are family varieties developed by a lone individual (and named after him or her) but eventually claimed as a family heirloom. It is not unusual for a gardener to work with a single variety for decades, trying to improve it through selection year after year. For instance, the Yellow German type has many variants of different sizes, ranging from one ounce to several pounds.

Three Willard Wynn Yellow German tomatoes coming along. (Dobree Adams)

Much of this variation comes from gardeners having a preference for a certain size tomato and saving the seeds from the plants producing that size. Some gardeners want a tomato that can be eaten whole at a meal, while others want large family-size slicers. In any case, the primary criteria for growers of heirloom tomatoes are flavor, texture, and disease resistance.

With the growing popularity of farmers' markets and CSAs (community-supported agriculture contracts between growers and consumers), we can expect heirloom tomatoes to account for an increasing share of the tomato market in the future. They are already popular enough to encourage major seed companies to offer them (sometimes reluctantly) in their catalogs to compete with small online heirloom seed distributors. CEOs of seed companies may refer to heirlooms disparagingly in major publications such as the *New York Times,* but that only makes them more popular. Heirloom tomatoes, like other heirloom fruits and vegetables, are here to stay.

5

Letting Traditions Speak

It is important to realize that heirloom fruits and vegetables raised by traditional growers almost always have a story behind them. These stories are important because they give a sense of how fruits and vegetables fit into the history and culture of a people at a given time. This is particularly true in Kentucky and particularly true of beans and tomatoes.

Kentuckians cherished their beans and took them wherever they went to obtain jobs during hard times. Most left with the intention of coming back, staying away only long enough to get ahead, to make enough money to pay off the house at home, to make enough to buy the homeplace to settle an estate, or to get out of debt. However, as statistics show, many never returned home except on weekends or to be buried in the family plot or high on a hill in the community cemetery, forever out of danger of being swept away in a flood.

On weekends in August, one can still find many gardeners parked along the roadsides in eastern and southeastern Kentucky selling fruits and vegetables, often together with flea market items. Many of the sellers set up at organized flea markets as well. The favorite garden items are beans and tomatoes. Many, if not most, of the buyers are transplanted Kentuckians on their way

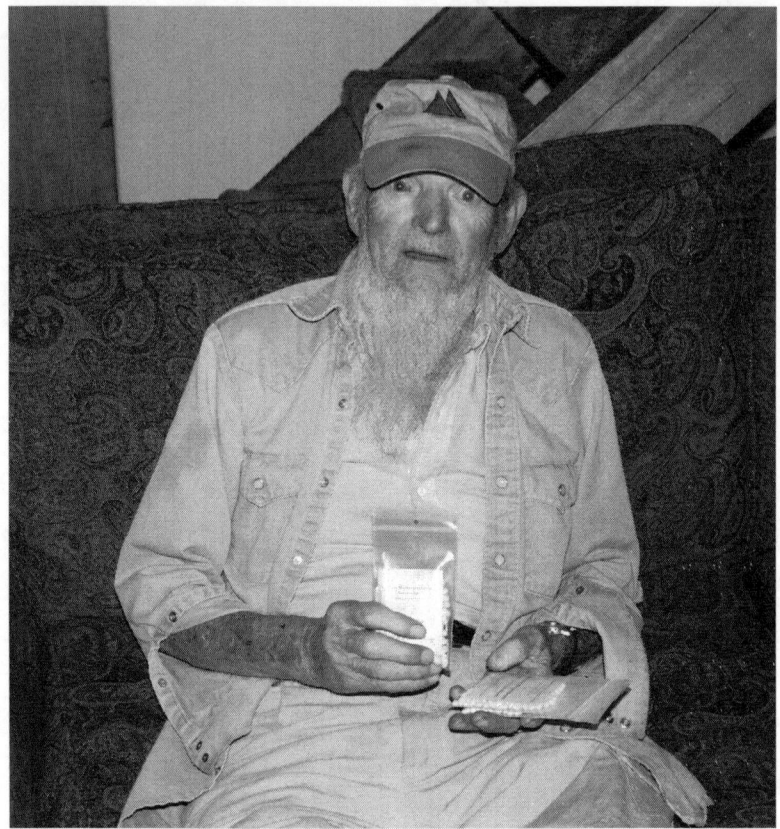

An old-time seed saver from Jackson County who stopped by to buy some seeds, perhaps to get a start, or maybe to share his seeds.

back to Ohio, Indiana, or Michigan. They stop to buy tomatoes and beans to eat fresh or to can or to make shuck beans. At one time, shuck beans were eaten on a daily basis, but now most people prepare them only for holidays such as Thanksgiving, Christmas, and New Year's Day or for important family occasions such as reunions, weddings, and anniversaries. Kentuckians who are now living throughout the country contact me each fall to buy shuck beans for all of the above-mentioned occasions.

Big John beans, a favorite in Letcher County, supposedly came to Kentucky around the time of the Revolutionary War.

Beans are often asked for by name. These names bring back memories of childhood, when almost everyone had a family bean that had been grown for generations. Big John beans, which are commonly grown in Harlan or Letcher County, often go for about ten times the wholesale price of a bushel of Blue Lake beans at a farmers' market. Blue Lake, the most plentiful of the machine-picked beans, can be found almost anywhere.

Many people heading back up north used to ask for half-runners, which were a favorite bean in the mountains until the seed companies exploited their popularity by mass-producing them in Idaho and harvesting them by machine. Unfortunately, machine-harvested seeds, grown by irrigation and not supported by poles or trellises, have little if any quality control, and genetic manipulation sometimes produces unwished for results. The modern commercial half-runner contains so many tough and inedible beans, typically more than half by actual measurement, that most serious gardeners won't grow them anymore. This has led to a resurgence of interest in the old-time beans such as the Big John, Paterge (Partridge) Head, Turkey Craw, Goose, Aunt Bet's, and any number of varieties of greasy beans and family beans. Additionally, the old-time half-runners are now being sought again and are poised to make a comeback, thanks to the dedication of many seed savers who are producing half-runner seeds in substantial numbers. These heirloom beans will continue to thrive only if enough growers are committed to saving the seeds from their beans—another tradition that was in danger of dying out but is now making a comeback.

One longtime agricultural supply store owner in Winchester, Kentucky, went to Idaho looking for someone to grow a popular greasy bean, only to be told that he would have to contract for a minimum of forty acres of beans grown for seeds. And, of course, there would still be the issue of quality control. He left without signing the contract, since there was no way he could sell forty acres worth of seeds, even for a very popular regional bean.

Handing Down Traditions

A few years ago, I was at a meeting of the Pine Mountain Settlement School's Board of Trustees. Several of us were sitting in a small informal group, engaged in a conversation about the importance of seed saving as more and more people were abandoning the practice and buying their seeds from catalogs. Several board members were practicing seed savers, which caught the attention of another member who taught sociology at Berea College. He had never heard of families having family beans that had been grown from generation to generation. He remarked that his family didn't have a family bean, and he wondered why such a thing even existed. He was in the company of several people with advanced educational degrees, many of whom grew family beans each summer; ate them fresh, canned them, pickled them, and dried them into leather britches (shuck beans); saved their own seeds from the best plants; and traded seeds with other gardeners. As the conversation continued, he realized that his colleagues were participating in traditions that had been handed down for centuries. Although these traditions had died out in much of the United States, they were still going strong in the southern Appalachians, where people were accustomed to fending for themselves in both good times and hard times—higher education notwithstanding.

Part of the irony of the situation was that the trustees of the Settlement School, as well as everyone else attending programs at the school, were being fed meals that included the most modern commercial beans and tomatoes, even though they were in the midst of one of the world's major strongholds of heirloom vegetables: Harlan County, Kentucky. The mountains of southeastern Kentucky have one of the most complete collections of heirloom beans anywhere in the United States, and a large number of heirloom tomato varieties as well.

One of the trustees was Dr. William Leach of Silver Spring, Maryland. He had maintained several family heirloom seeds, and his great-grandfather, William Creech, had given the original tract of land on which the Settlement School was built. The school also possessed a farm consisting of several hundred acres, much of it flat or gently sloping land that would be ideal for the production of heirloom fruits and vegetables. But "progress" had come to the valley, and even the Settlement School had become its willing victim.

The Role of Culture

What I am very aware of in my own extended family, and what I have increasingly observed among other families in the southern Appalachians, is the role of cultural traditions in the continuation of seed-saving practices. Because many who migrated to this country settled in areas that had already been urbanized, they immediately became dependent on the sellers of food products. They had no reason to continue or develop traditions of self-sufficiency and always worked for the "other man." In contrast, many Appalachian settlers came to the United States to escape religious or political persecution and quickly became self-sufficient, depending on their crops, their animals, and foods from their natural surroundings. They learned from the Native Americans which herbs to use for medicines and which wild fruits and berries were edible. Of course, they were also quite dependent on game animals in the early settlement period.

To get the necessary tasks done in a timely fashion, extended families worked together in planting, growing, harvesting, and preparing crops for winter use. There was no such thing as a family garden. There were numerous gardens, scattered in many areas, to take advantage of even the smallest patches of fertile soil. Many families also had at least one garden near the main road to share with strangers passing through who might be hungry. My maternal grandmother had a garden adjacent to the road in front of her house for just that reason.

When someone in the family became ill or incapacitated, the whole community might descend on the affected family and plant, cultivate, or harvest the entire crop. Or someone might stay with the family and take on most of the chores. When my father had an appendix operation followed by a rupture operation, my first cousin, who lived more than a hundred miles away, came to live with us during the summer and quickly learned how to plow with mules. He took over most of my father's roles until he was "up and running" again. This cousin was about five years older than I was, probably thirteen or fourteen. In other words, helping out was just what one did, and no one gave it a second thought.

The most graphic illustration of helping out that I can remember happened when I was about seven or eight years old. A longtime bachelor in our community was gradually going blind and becoming unable to maintain his small farm and fend for himself. He had always been totally self-sufficient and never would have asked for help, but everyone knew he wouldn't survive without it. Because most of the people in our community were related to one another, either closely or distantly, he was also kin. Cousin Hermas McCracken was in need, and the community rallied. When Hermas became confined to his home, people brought him food and made sure he had plenty of wood stacked next to the fireplace. He would never eat in the presence of anyone, but his neighbors made sure he always had plenty of food within easy reach. When he became too blind to find his way to the spring, some of the men rigged a wire that he could hold on to and walk to the spring to get a bucket of water. I remember going with my father to Hermas's house; we brought him some food, checked out his situation, and stayed and talked for a while. When he finally became unable to meet his most basic needs, even with the community's help, Hermas was taken to the county home, where he spent his last few months. People from the community continued to visit him there, taking some of his favorite foods. My first cousin even brought his banjo and played Hermas's favorite music—music he had danced to as a young man.

It is this same spirit of community that has led to the survival of so many heirloom seeds in the mountains. Everyone knew the importance of saving seeds for the next growing season. When you have something good, and you know it's good, you hold on to it, even if its value isn't always readily apparent. Only the most irresponsible person would "eat his seed corn."

The Gallahar Bean

The story of the Gallahar bean arrived through e-mail from Minnesota, but the bean originally came from Tennessee. This seed saver preferred to remain anonymous, but the story of her family bean is all about tradition.

> I was worried that I was being presumptions [*sic*] with this old bean and didn't want to waste your time. Remember though, this is the "Gallahar" bean, if you find it is a "unique" bean and want to name it. I was afraid that this bean was going to die out . . . cousins and others have been given seeds over the years, but like me have not "kept seed." (Our springs are cold and wet up here, not like Tennessee.) Finally time has caught up with me and dumb as it will sound to many, I couldn't stand the thought of it being gone forever. I found your site when I Googled "Heirloom Beans." I was trying to find seed for a "peanut bean" that my aunts used to grow. Appreciated your interest in "old seeds" but it took a while for me to get the courage to contact you . . . so glad that I did!!!
>
> I also have a very limited amount of mustard seed that my grandmother, then my aunts (her daughters) grew, if you are interested. This mustard is/was used for greens, not for spice seeds, no name or history—just 100+ years old, hopefully unpolluted seed.

She sent me the bean seeds on August 24, 2010, accompanied by the following explanation:

Enclosed are the pole beans we spoke of yesterday. The Gallahar, as they have been known since my great-grandparents were given the bean. The spelling is my guess as the name was never written down. The Gallahar was being grown when my grandfather was a child; I don't know the real age of the bean.

My grandfather was a farmer and grew acres of this bean in the corn fields on the bottom land of the Clinch River, selling at market in Knoxville, Tennessee. This stock is what I grew from last year for my aunt. I got about 90% germination even though I was growing in Minnesota (cold wet spring) and this stock is from 2001—and not stored in a freezer.

In case you have a Liles or Yarnell bean, it might be the same as this one. Thank you for your interest in this old bean. I hope it isn't a waste of your time. And a huge thank you for your devoted ideology in maintaining the heirloom seeds bank.

After some phone conversations about her family's work with the bean, she sent me this historical account of her family and the family bean:

Upon being asked to write a story to go along with and document THE bean, my first reaction was panic—& forget that—then upon being asked a second time, my mind went to a soap box opportunity here. I could blast the Monsanto's, the gene manipulators, the Frankenstein makers of our gardens, but I know what Mr. Best wanted was a history of the bean that I had given him (with much trepidation and an apology for having the audacity to ask if he would want that old bean!). I had found Mr. Best when

Googling "Heirloom Beans" and immediately felt that he was a keeper of the past as a gift to the future. So for him I will try to write the story of the Gallahar bean.

THE BEAN, called the Gallahar, has been passed thru five if not more generations. My grandfather, WW, received the bean from his father Richard H. Yarnell (Richard's father was from North Carolina, with his father from Pennsylvania) and mother Rachel Liles of Anderson County in the east Tennessee hills. Grandpapa thought the name Gallahar was more than likely the name of the person that gave the bean to his father or his grandfather, as the bean was around long before my grandfather.

WW was, among other things, a farmer, as was his father before him. Farming was a given, and sustainability was the natural, normal way of life back then, and you kept your seed from this year's garden for the next year if you planned on eating the next year. For my grandfather, the Gallahar put food on the table in more ways than one as he grew the twining bean in the Country Gentleman and Hickory King cornfields they planted in the bottom land on the Clinch River. The bushels of produce from those fields were taken "to market" to sell in Knoxville by black horse and wagon, with the horse allowing my grandfather to sleep on the way back—she knew the way home better than my grandfather!

My memories of the bean fields are of the heat, the hot, hot sun—stinging packsaddles that got you—ground too hard to hoe—my grandfather fishing on the bank alongside those fields—of family. After my grandfather no longer "farmed," my aunt grew the Gallahar in their garden on river cane poles. Then after my grandfather passed on, two of my aunts moved back with my aunt that had always "stayed at the home place."

They continued keeping seed year to year, sometimes selling beans to those in the area that knew of their garden and would stop by and ask—canning most and giving a lot to family. There still sits on the shelf back home the last quart of Gallahars from the aunts' last run of canning beans. My personal opinion of this bean is that it is far better as a canned bean than fresh or frozen; it is a string bean—gets fibrous if left on the vine too long—but my aunts always depended on this bean as the reliable producer for "putting up" for the winter. Each winter the new seed catalogs would pour in and the three sisters would "try" a new variety but they always had the Gallahar in their garden. Although I try to garden and "keep seed" as I was taught, I was most grateful that Mr. Best wanted to have a sample as it seemed "wrong" to let this 150+ year old bean fall by the wayside and into oblivion—and he is the keeper of heirlooms.

Aunt Bet's Bean

About thirty years ago, Bill Hayes, a fellow trustee at the Pine Mountain Settlement School, gave me some seeds of Aunt Bet's bean. The bean had been grown in the Hayes family for several generations, and he was anxious to share the seeds with me, as I had also shared some of my heirloom beans with him. I had good luck growing Aunt Bet's bean, and it has been part of my collection ever since.

Fast-forward thirty years. I talked on the telephone with Calvin Sammons of Bartlett, Tennessee, and then received this letter from him on November 10, 2009:

Enclosed are the Bet's (Aunt Bet's) beans we discussed along with several other pieces of information we wish to share.

When we were talking about seeds I should have told you there was one plant of which I have been searching with no success. When I was a child growing up on the side of a mountain in West Virginia, my grandmother Bessie Hopkins Sammons was the gardener on the steep hillside of our land. All the normal crops, but one was different from all her neighbors, and that is what I have been searching for.

She called them muskmelons. To her the round shaped melon with yellow insides was cantaloupes. Muskmelons were yellow inside but they were shaped differently. They were long and tubular with a fat end. If you can remember the shape of the caveman Alley Oops' club, you would get the idea. I remember them as very sweet, a strong-sweet smell, and thin skinned.

The ones I found once upon a time from a seed catalog just were awful tasting. Tasted more like a gourd than a cantaloupe taste.

Anyway, if you ever run up on these seeds, keep me in mind.

Along with this letter he sent me some genealogical information dating back to 1717. Aunt Bet's bean didn't become part of the family until the early 1800s.

Calvin's wife, Waukesha Lowe Sammons, also attached a note detailing her family bean's connection to the Pine Mountain Settlement School:

The kinship [below] could be the reason my Brashear family heirloom pole beans were being grown on Pine Mountain and at the Pine Mountain Settlement School.

Waukesha Lowe Sammons
Great-great-granddaughter of Eli Brashear and Sarah
"Sally" Campbell Brashear
Great-great-great-granddaughter of Elizabeth "Betsy"
Young Brashear (1808–1902) and James N. Brashear, Sr.

PINE MOUNTAIN SETTLEMENT SCHOOL
Harlan County, Kentucky—A National Historic Landmark
Eli Brashear wed Sarah "Sally" Campbell.
Eli's mother: Elizabeth "Betsy" Young Brashear,
progenitor of our heritage pole beans, "Betsy's Beans."
Sarah's sister, Mary Campbell, wed Joseph Creech.
Mary and Joseph's son, William Creech (1845–1918)
wed Sarah "Sally" Dixon.
William Creech donated land to establish the
Pine Mountain Settlement School.

At the end of the genealogical information, Waukesha and
Calvin Sammons added this note:

We have worked hard to keep the beans from being cross-
pollinated by never planting them close to other beans.
Otherwise, next year's seed beans will not produce the
same type of crop of beans. The beans are excellent fresh,
frozen, canned, or dried for shucky beans. Always save
enough bean seeds for planting the next year. Please
help sustain the history and the heritage seeds for future
generations. By remembering the past, we honor the
future. With the sharing of our heirloom bean seeds,
we extend our gratitude to all who participate in the
protection of our Appalachian heritage.

Rescuing the Noble Bean

I receive e-mails and phone calls every day about heirloom beans, but Judy Bennett's letter of June 27, 2008, from Dayton, Oregon, was different from most:

> I am trying to locate a bean that was grown in West Virginia. My great-grandfather brought the bean with him when he moved to Oregon in the late 1890s. He and my grandmother grew it in the garden here and my grandmother and mother always canned the bean. I just recently got some of the seed from my great-aunt (who had gotten the seed from my grandmother), but the seed has been sitting in her garage for about twelve years. We soaked the seed, it enlarged, and planted it, but it isn't coming up. The seed is brown (I still have some if I could mail it to you to look at) or I could take a picture and e-mail it. From what I remember, it grew really big in the hull (filled it out) and was picked when the hull was starting to turn a light yellow color.
>
> You canned it in the hull but after pressure cooking it, some of the beans would be out of the hull and some stay[ed] in. The hull turned a brown color and was really tender to eat. My great-aunt said that my grandmother and great-grandfather just let the beans spread out on the ground, but when she and my great-uncle raised it, they would string them to grow up.
>
> Any help would be appreciated and any ideas about planting the seeds I do have. They are good and dry.
>
> Thank you.
>
> By the way, we called them the Noble Bean, as that was our last name. My great-grandfather came from Walton (Roane County), West Virginia.

At first, I thought she might be talking about the Logan Giant, which is a well-known bean in West Virginia. But when she sent me photos of the beans, I realized that the Noble was not the same as the Logan Giant.

We continued to exchange e-mails, and Mrs. Bennett decided to send me several hundred beans to see if I might be successful in getting any of them to germinate. Upon receiving the beans, I immediately realized several things:

1. The beans had been dampened enough at some point to swell up, and then they had dried out again.
2. The eyes of the beans were split, and the embryonic beans had apparently dropped out of many, if not most, of the seeds, which would make germinating them much more difficult. In fact, I didn't find any seeds that I thought might germinate.
3. I was in for a challenge, but I decided to give it my best try.

Using some greenhouse starting soil, I put about a hundred of the bean seeds in a flat and moistened the soil enough to keep the seeds damp. I put the flat on top of a bench on a sunny porch so that I could inspect it several times a day and keep the moisture at the appropriate level.

After three weeks, I was becoming more than a little concerned that none of the seeds would sprout, but then my wife noticed a small amount of green near the middle of the tray. Shortly thereafter, six young beans broke through the soil, leaving me quite relieved. However, a grasshopper quickly found one of the beans and ate it, and I realized that some precautions were in order. With success almost at hand, I didn't want to take any chances. I transplanted the five remaining plants into pots and put them in my greenhouse, where I allowed them to develop roots prior to transplanting them in one of my two high tunnels. (By that time, it was too late to transplant them in the field, since they could not possibly produce seeds before freezing weather arrived.)

On October 5, 2008, I e-mailed Mrs. Bennett the following information in part: "One grew really well and three are still struggling. The fifth succumbed to damping off. I believe the one that is growing well will produce at least a dozen pods of mature beans from the way it looks now." As it turned out, I was a little too optimistic. Three of the remaining four succumbed to attacks from ants and dung beetles, and the one healthy plant produced only four pods of viable seeds before succumbing itself. But the Noble bean was now twelve seeds from extinction.

I notified Mrs. Bennett of my luck on October 26 and told her that I would soon be sharing the seeds with her. I sent five to her and gave one each to John Coykendall and Lothar Baumann. John Coykendall works with heirloom fruits and vegetables at the Blackberry Farm near Maryville, Tennessee, and is a noted seed saver. Lothar Baumann is a truck farmer near Berea, Kentucky, who is also an heirloom bean collector and grower. Of the five remaining seeds, I decided to plant three myself during the summer of 2009. I kept two to try again in case we all experienced crop failure.

The summer of 2009 was good for the Noble bean. John Coykendall grew his plant indoors and had excellent success, with his one plant producing more than 380 seeds, half of which he sent to me. Judy Bennett also had good success with her five seeds, yielding enough to plant several rows in her garden in 2010. My three plants, planted in my greenhouse, produced around 100 seeds, giving me enough to plant in the field in 2010 and to share some with Frank Barnett.

Judy Bennett had excellent success with her 2010 crop of

Noble beans. In addition to the many meals cooked from her crop, she canned more than fifty quarts of beans. Frank Barnett also had good success with his few seeds and had plenty of seeds for another crop the next year. I planted some of mine on two different occasions and now have enough seeds to grow the Noble bean to sell at farmers' markets and to sell seeds to the public.

The Noble bean is what is commonly known in the southern Appalachians as a fall or October bean and is an outstanding example of the type. It is stringless and tender. I believe it is now safe from extinction, and I'm glad I played a part in bringing it back. It was later determined that the seeds had been in the garage for seventeen years, not twelve!

Melody Rose

Melody Rose grew up not in Appalachia but in western Kentucky. But she has the same vivid memories of the family vegetable garden:

As a seventh generation Kentuckian, I'm acutely aware of history and tradition. My ancestors came down the Ohio River in 1775 with General George Rogers Clark and were among the first families to settle Corn Island, which ultimately became Louisville. I'm fortunate to have a family that respects history and have also been avid gardeners.

I grew up in Marshall County, western Kentucky, and some of my earliest memories were of feeling the freshly turned soil under my bare feet as I dropped bean seeds in the rows for our vegetable garden. Yellow summer crooknecks, speckled horticultural beans, purple hull peas, and lima beans were always a part of our 1960s garden. I especially loved harvesttime, and summer was filled with a house full of aunts breaking beans, scalding tomatoes, and canning produce from the monstrous garden that covered a couple of acres.

Flash-forward several decades and I now had a home and property of my own and room enough for a large garden. My first endeavors were lackluster, to say the least. I needed to hone my gardening skills and also discovered that the vegetables didn't taste like I remembered.

By doing research on the Internet, I discovered a whole culture of seed savers who were searching for old varieties no longer in commercial production, and I was hooked. Generous traders sent me some seeds for postage, and since I wanted to contribute too, I started the hunt here in west Kentucky. I started asking friends, family, and neighbors whether they knew anyone who might be saving vegetable seeds.

My first "find" came from a friend of my husband's, Danny Mullins, who lived in neighboring Graves County. His wife Dana sent me what the family called "stock peas." She was also gracious enough to share the family history that went with them. In keeping with common seed-saving traditions, these seeds were handed down through the female line of the family. We can trace these peas back to her great-grandmother, Pemealier Nall, born in 1862. She handed them down to her daughter, Bertie Nall Adair, born in 1890. Dana's great-aunt Evelyn Adair Rodgers, born in 1909, shared them with her. The family settled in Graves County in the late 1820s and most likely brought the seeds with them, but we do not have documented evidence of such.

By that time, I was intrigued to know if these "stock peas" had an actual name, so Brook Elliott sent a sample to William Woys Weaver to see if he could identify them. His report came back stating that these were a very old strain of

Whippoorwill pea. This experience just encouraged me to continue the hunt here in west Kentucky for more of these quiet seed savers.

A former high school teacher offered seeds of a large pink tomato that her great-uncle brought from Germany in 1919. Nancy Forsythe told me how her great-uncle Mark Bagby (possibly Bagbhe) arrived in west Kentucky with his tomato seeds, and the family has been growing and sharing them ever since. This is a large, pink, potato-leaf tomato that does well in our hot, humid summers. Nancy and her family were very generous with these seeds, and on numerous occasions I discovered it being grown by unrelated families. Since she was a teacher, Nancy shared the seeds with students and colleagues, making Uncle Mark's tomatoes beloved by many. Often it was known by other family names, but when I dug further, most of those families could trace the tomatoes back to a former teacher in the Marshall County school system. Even my own uncle Joe Hall proudly presented me with tomato seeds that he called Dr. Ford's Tomato. However, since I knew that Dr. Ford's wife, Beverly, taught school with Nancy, it wasn't hard to connect the dots back to Uncle Mark Bagby.

The many incarnations of this tomato taught me a valuable lesson in following through with additional questions and research, because what appears to be a new variety could simply be a renamed familiar face.

Melody Rose, together with Brook Elliott and Gary Perkins, started the Appalachian Heirloom Seed Conservancy, which lasted for a few years. The annual meetings were held in the barn at my farm.

Letters Tell Part of the Story

Letters arrived in droves after that first story about my family farm appeared in the *Rural Kentuckian* in 1988. The following excerpts tell over and over again the importance of tradition and culture behind the simple act of saving seeds from one year to another.

Opal Cummings, Frankfort, Kentucky

The article in the *Rural Kentuckian* mentioned my family, which included my wife, Irmgard, and our two youngest children, Michael and Barbara. Like this one from Opal Cummings, many of the letters referred to my family, and some were addressed to my wife and me:

> I am interested in the "greasy beans" and wonder if you would sell me enough seeds to get a start of them. I am from Menifee County and I remember when I was home, my mother raised these beans and she called them "greasy grits." They were a delicious bean and so much better than the White Half-Runner that everyone wants now.
>
> I have been away from home now for 38 years and my parents are both gone and your article is the first I have seen about the "greasy beans." Whenever I mentioned them no one knew what I was talking about, so I was quite excited when I read your article.
>
> If you have extra seed that you can let me have, just put a bill in with them and I will be happy to send a check. Your son and daughter are fortunate to be part of a family that appreciates gardening. I think it is a wonderful feeling to see the results of the little seeds you put in the ground. Plus being able to feed your family. It does make

the family closer to work together. It's good therapy and a wonderful way to relax after a bad day at work.

It is interesting to note that greasy beans are still being grown in Menifee County, and Frank Barnett found some at a farmers' market. It is entirely possible that a greasy bean I got from Frank and grew in 2010 is the same greasy bean Opal Cummings knew as a child.

Gude Henrick, Hardinsburg, Kentucky

I also am a Gardener but on a very much smaller scale than your operation. Mine consists of about one acre, producing for the household plus hoping to have some extra for sale. The last two years due to small rainfall has been unfortunate. However, I will try again I feel sure.

I am interested in securing two of the products that you mentioned in the article—namely "greasy beans" and "German pink tomatoes."

Will you please sell me enough seed for a start with these two? Also where do you buy your irrigation hose? Your assistance will be much appreciated. I wish you a bountiful year.

Mr. Henrick's letter mirrors many others—expressing interest, comparing notes, asking questions about where to obtain supplies, and hoping to buy or trade a few bean or tomato seeds to get a start.

"Getting a start" is a well-known phrase throughout Kentucky and the southern Appalachians. I use it myself, often offering to exchange seeds to get a start with a particular variety while doing the same for someone else. It is possible to get a start with only three or four seeds, and I once drove more than 400 miles round-trip to pick up six seeds near Chattanooga, Tennessee. Unfortunately, I have had two to ten seeds sent through the mail smashed by the stamp-canceling machine. Now I always pay the extra cost to mail bean seeds in padded envelopes and have them stamped by hand.

Mrs. George Washum, Walton, Kentucky

I used to have the German Yellow and German Pink tomatoes but over the years I have lost them.

I was wondering if I could purchase just a few of these seeds to get myself started back. I do not raise vegetables to sell, only my own home use and canning. Also I would like to purchase a few "greasy bean" seeds. I have heard so much about them but am unable to find them.

Two weeks later, she wrote to thank me for the seeds and asked to be put on a list to get seeds the following year.

Laura D. Sutton, Willisburg, Kentucky

I sent bean seeds to this lady in October 1988, and her letter came the following spring:

Here are some cornfield bean seed that have been in my family for a long time. My grandmother raised them many years ago. She died when she was eighty-eight, and has been dead for forty-six years. My mother gave me the seed and I have been growing them for fifty-six years, so you see I am no youngster.

I planted my greasy beans today. I can't stick them, but have a piece of woven wire fence that I will let them run up on. Hope I have good luck and I thank you for the seed. Hope we have a better growing season than last year.

Her letter illustrates the age of many heirloom beans that have been in families for generations. Her grandmother probably got the beans when she was in her late teens, and they had probably been raised by her mother and grandmother before her. We can easily go back 150 years and more.

Viola Bible

I just finished reading the article in the July 1988 [issue] of *The Rural Kentuckian* about your family farm. I enjoyed it very much. I was especially interested in the fact that you were growing greasy beans in your garden. My husband and I are trying our hand at gardening since we retired and moved here from Michigan. The first thing I wanted to try was greasy beans, but all efforts at finding seed have been to no avail. I was hoping that perhaps you might have a few extra you would share with me. My mother who is eighty years of age grew them when I was small in eastern Kentucky. She lost the seed several years ago and has been trying to find them also. I would appreciate it if you could just spare a small handful. Enclosed is $5.00 to cover handling and postage. The self-addressed envelope is enclosed in case you can't spare any [but] you might know where I can get some.

Sadly, we don't know where the Bibles moved after they retired and left Michigan. Her letter illustrates many things about seed saving: culture, migration and return, and a hunger for foods eaten as a child. It also illustrates that heirloom seeds never became part of the seed industry that dominated American gardening after the 1940s. If you wanted to have the vegetables of your youth, you had to save seeds or, if they became lost, find them again from some source, usually from kinfolks. With my own seed saving and the attendant publicity, I was becoming kinfolks to many people, without intending to do so.

Mrs. Kenneth Gross, Harlan County, Kentucky

I hope this will reach you. I read about you all in the *Kentuckian* and I would love to have some greasy bean seed. Be glad to pay for them. Also I'd like to know where the market is that you go to sell your vegetables. I'm a gardener too and belong to the Seed Savers. I've got some good bean seed too. I really enjoyed reading the write up. I made all my eight kids farmers, too.

This short letter, which was addressed to "Bill Best and Family," said a lot in a few words. It was obvious that Mrs. Gross took her gardening seriously, as she belonged to the Seed Savers Exchange, an organization many Kentuckians have joined over the years. She also "made farmers" of her eight kids. All children were expected to help out in the family garden from an early age. They might not have always liked it, but most realized that the family depended on their work.

Another article about my family's farming operation appeared in the *Lexington Herald-Leader* on August 29, 1990. Although it didn't bring as many letters and requests as the earlier article, it did focus more attention on our work with heirloom beans and tomatoes.

Claude H. Brown, Pikeville, Kentucky

Dear Fellow Hobby Gardener!
I have in front of me the clipping from the *Lexington Leader* of 8-29-1990 which prompts this contact with a "Colleague of the Garden." You indicated in the clipping

that you could match the Heinz varieties if you could find two more tomato varieties. The enclosed exclusive Brown varieties will help you achieve your goal. According to the clipping, I am approximately 28 years your senior which means my hobby experience and the joy of gardening exceeds you.

I have developed and controlled the up-grading of these three Brown identifications since my return to my "fanatical" hobby after my military discharge in 1945. The Brown's Yellow Giant tomato is definitely my own "concoction." I am offering to you the "exclusive rights" to propagate these varieties and display and sell the products at your Farmers' Market. A favorite friend near Pikeville sells these tomatoes in his side road market for a straight $1.00 each—and cannot meet the demand. Since I am V.P. of the Board of Directors of the U.K. Pike Co. Extension Board, I am in the position of recommendation of these varieties.

I am also extending to you (since you are a challenging colleague) the opportunity of growing several hills of the Brown heirloom cushaw. Thus, I am enclosing seed of the Old Fashioned Solid White Gooseneck Cushaw—the seed of which has been "handed down" thru the Brown family for over 75 years. The Pike County Farmers' Mkt. sells my Goosenecks for $6.00 each. You should have seed from a perfect specimen you can grow from this enclosure. Since you will be the only person growing them, I assure you this display will be a real "attraction getter."

I am enclosing certain "Suggestions" for successful culture of my tomatoes and Cushaw. You will add your own additions or changes to ensure success in 1991. Since

I visit often with my banker son in Lexington, I will visit the Mkt. there this summer in the hope of seeing you. I am a retiree of Pikeville College which means you and I have a mutual love for the schoolroom.

Best wishes in 1991 for health, wealth, and happiness.

Mr. Brown wrote another letter on April 24, 1991, which I received as I was planting my gardens that summer.

Hello Gardening Colleague!

You and I share the same mutual hobby—gardening. Seems you are primarily a "bean man." I specialize in tomatoes. However, I assure you I'll do my best with the beans you sent.

I am sending you two samples of beans that are considered favorites here in E. Ky., Va., and W. Va.

First, the State Half-Runner. Our women folks prefer this bean because it looks so nice in the can, has a large white bean that enhances its eating quality, and cooks to a golden brown and tender while preparing for the table. U.K. Ag. Extension uses it primarily as a salable bean at our local Farmers' Market. I trellis my State Half-Runners and it becomes a beautiful Early Six Weeks bean when picked. The second bean I am sending is the Early Six Weeks bean. It's a bush bean that is the first after-frost bean you can plant. I have already planted a 100 foot row across my garden. It is the earliest bean you plant in the spring garden. You will find that the Six Weeks is the first bean ready for the market.

Write me this fall after you have produced my three famous tomatoes and the heirloom Solid White Gooseneck Cushaw. Give me your successful report.

Fraternally,
"Brown"

P.S. The State Half-Runner should be trellised or grown in corn.

Claude Brown was certainly correct when he spoke of the superiority of his Brown's Yellow Giant tomato. It is one of the best of the deep yellow tomatoes. He also sent me seeds of the Mortgage Lifter, which was developed by Radiator Charlie in West Virginia, near Mr. Brown's home. I had already grown that excellent tomato.

He mentioned that he was a member of the Seed Savers and had sent his squash seeds to growers in all fifty states. He had apparently sent his yellow tomato seeds to many people as well, since I found that his Brown's Yellow Giant was being sold on the Internet by several small heirloom seed operations without giving him credit. We have been selling heirloom tomato seeds as well as bean seeds on our website for a number of years now. I call his tomato "Claude Brown's Yellow Giant" and give him credit for developing this fine tomato. I know of no other sites that mention his name. I never got to meet Mr. Brown in person, but his tomato is one of the best of the heirlooms, and I applaud him for taking the time to develop a tomato for its disease resistance, texture, and flavor, rather than a long shelf life and the ability to be shipped thousands of miles.

Letters from Afar

Through the generations, Kentucky culture and seeds have traveled far and wide. Memories of plants that children knew when growing up tend to stay with them throughout their lives. Seed saving is a vital part of Kentucky culture, and even when people have moved elsewhere, they often go to great lengths to recover seeds lost for one reason or another.

Ron Black, Raymond, Washington

All of my family is from Letcher County, Kentucky (Mayking and Whitesburg), and Wise County, Virginia, near Pound (Flatgap). My family settled the area and has lived there over 250 years so our mountain roots run deep.

My parents brought us to this small town in Washington back in 1964 when I was eight years old, now 54. My grandfather owned a small coal mine in Mayking and my mother was determined to not have us work coal mines. To my surprise there were about a dozen families living here in Raymond, Washington, who had moved from Wise County, Virginia, and Letcher County, Kentucky, all working here as loggers.

Half-runners will grow here and we are always on the lookout for a good white half-runner bean. We grow them for the bean but all agree it is to preserve some family traditions of growing this type of bean as our grandparents did.

So I am excited to get these seeds and will share a few with family and friends; hopefully they will like them enough to order more from you folks. Recent purchases of seeds from other growers in Ohio have been tough and non-edible. My garden is only 20 by 30 so I don't have much room for growing inedible beans.

Thank you very much.

Ron (Holbrook-Sturdily) Black

P.S. My late grandmother Sturgill was born in Lee County, Virginia, and she loved white half-runners. She would always talk about them come summer and these stories were wonderful. I see you sell a preacher's bean from that area. Is it of good quality and not tough? Maybe I will try them next time next year.

Deb Schroer, Cambridge City, Indiana

Can't believe I found your website. My mother's family was from Hazard, Kentucky, my father from Wooton's Creek (I believe in the Perry County area). Beans were always a staple in our family garden growing up. To this day, there is no favorite vegetable than beans, tomatoes, and corn. Question . . . a few months ago I purchased decorative bean plants called caracalla beans. Have you heard of this type of bean? My family were constantly drying seeds. Mother's favorite place was on an old screen door in the attic. When dried, she would put the seeds in an old feed sack. I love those memories. Great to reconnect with someone from Kentucky. I always say the nicest people are from eastern Kentucky.
Look forward to hearing from you.
Deb Bailey Schroer

I responded to her question as follows:

Hello Deb,
According to information on the Internet, the Caracalla bean is sometimes called the snail bean and is considered inedible. The Goose Bean, found widely in eastern Kentucky and throughout the southern Appalachians, is one of the favorite beans of the region and is still widely grown. Thanks for the information on your family's love of beans and your mother's techniques for drying them. If you would like to order some of our beans for this

summer, you still have plenty of time to grow them for making shuck beans this fall.

Deb's next letter shared even more childhood memories:

Thank you for your reply. I will send an order this week. One of my favorite summer memories is breaking beans with Granny. Mother always "put up" at least 500 quarts of White Half-Runners. Having Granny there and the stories she told always made the time go by much quicker. I cannot remember not being a part of picking beans (in Indiana they call it harvesting). I do remember being quite proud that my job was standing on a chair at the kitchen sink washing canning jars. I had the smallest hands, which had plenty of room washing every little cranny spotless.

I can close my eyes and remember the smell of freshly broken beans and the unique smell of cooked beans under pressure in the large cooker on the stove. Mother was quite proud of having only one or two cans that would not seal. Those 500 quarts of green beans were always shared with neighbors, family, and friends (our nuclear family as it is referred to now consisted of Mother, Daddy, my brother, and I).

I recall the first time I ever ate beans not home grown and canned. I was at an office meeting of some sort in a restaurant. I took a bite and had to force myself to swallow them. The look of the beans on my plate was different, and to this day I recall that instance when I realized my Kentucky roots had distinct advantages from other lifestyles.

Thought you would enjoy those stories.

Sharon Johnson Williams

I thought that I would type this for you; that way you
wouldn't have to try and read my chicken scratch.
My mother's name is Devella Johnson. Her maiden name
was Neal. She is the daughter of Harvey Riddle Neal and
Eliza Sparks Neal. My mom is one of five daughters—
Devella, Sada, Mary, Laura, and Lois. My grandfather
was named after a family that his mother worked for
by the last name of Riddle. Her name was Susan Neal.
My grandmother was the daughter of John Alden and
Elizabeth Frances Sparks. She was one of eighteen
children (his first wife died) and she had a twin sister
named Rosa who was always called Rosie.

My mother and her family moved to Indiana in 1936.
After talking with you the other night, I talked more
with my mom and she thinks that she was 15 instead of
13 when they moved to Indiana. She was born in 1921 in
Estill County, Kentucky, in the town of South Irvine.

When they moved to Indiana my grandmother brought
her garden seeds with her. She always planted beans that
she called a brown bunch bean, Rosie Beans (called that
because her sister Rosie gave them to her), cut-shorts, and
greasy beans. She also brought seed for Hickory Cane
corn. She planted that every year because my grandfather
would only eat cornbread made from that corn. They
made a trip to Kentucky once a year to the home of
Vernon and Provie Elliot (a relative) in West Irvine. He
had a grist mill and would make the cornmeal for them.
Grandmother would tie a blue ribbon or cloth around the
last three plants of each row. These plants were not to be
touched by anyone but her; those were her seed plants
for next year's garden. She fell and broke her hip in 1956

but was back in the garden the following year. She had a garden every year. She passed away in the fall of 1976 but had already started drying apples for her dried apple pies. Those were the best tasting pies I have ever had. (And I've had quite a few pies!)

I have never tasted green beans or cornbread that tasted like Grandmother's. The smell and taste of those beans was wonderful. They were so tender and flavorful. Nothing that you can buy in a store compares to those. I have bought garden seed up here and it is just not the same. I am so excited about having good old Kentucky beans again. So is my mom; it has really lifted her spirits.

Liz Smathers-Shaw, Shade, Ohio

The following letter from the area of my own family's roots was a special delight:

To refresh your memory, I'm one of the musical Smathers family from Haywood County, and am now transplanted to southeastern Ohio. My sister June taught Tracy Best how to play the banjo, and I played many a tune with Carroll (Best) years ago. (I'm a fiddle player.)

I have a real hankering to put up a lot of leather britches this summer. I'd like to try several varieties that you would recommend. My sister, who is the manager of the greenhouse at the North Carolina Arboretum in Asheville, and I cannot remember what our mother used to grow. What we do remember is how she put up bushels of leather britches, and not by stringing them, but by drying them on old window screens out in the sun, and moving them into the basement if the weather turned on her. One time she dried some on a screen in the back of

Hollie Wickers, niece of Clova Collins, looking over some seeds from her aunt's heirloom bean collection. Clova was one of my regular customers at the Lexington Farmers' Market for many years until her death well into her nineties. Some years later Hollie, in her mid-eighties herself, searched me out and gave me some heirloom seeds from her aunt's collection. I think Clova was originally from Floyd County, Frank Barnett's home county.

our car with the windows rolled up! We were very serious about our leather britches, and I'm so sorry we didn't keep the beans that she used, or at least remember their names.

I know that she liked cornfield beans but canned something else, so I'm thinking leather britches were some sort of cornfield bean. I also know that she didn't like to can "greasy" beans for whatever reason, and she only froze a certain variety which I can't remember. (She hated Blue Lakes, by the way, and she said they reminded her of growing beans for the Stokely company when she was growing up in Madison County. I guess it wasn't a pleasant memory!)

So, I'm hoping you can make some recommendations, and I'll get an order right out to you!

One of my favorite fiddle tunes is "Leather Britches," by the way!

6

Growing, Eating, and Sharing

Frank Barnett's Stories

Frank Barnett, a gardener all his life, has been actively collecting and sharing the beans of his youth since his retirement. He was born in Floyd County, Kentucky, and spent most of his childhood there. He speaks the language of the people in eastern Kentucky who are still holding on to the gardening and seed-sharing ways of their ancestors and kin. For the most part, I'll let Frank tell the tale in his own words.

Frank's Family

A large portion of my childhood was spent in Columbus, Ohio. My father worked for the C&O Railroad at the Roundhouse in Columbus, which maintained steam and later diesel engines. He finally got transferred to the Race-land car shop, where coal hauler cars were built in Greenup County, Kentucky, when I started high school. My sister and I were glad to leave the shotgun house on the south

Grandma Ella Barnett, late summer, in the sweet corn patch, with greens in the foreground. (Frank Barnett)

side of Columbus and move back up a hollow in Floyd County. It was a big improvement in our lives.

We didn't have much space for a garden in Columbus, but there were a large number of roadside markets in the countryside. A large population with Appalachian roots meant that White Half-Runners were the predominant bean sold in the 1950s and early 1960s. Canning was a big priority, and usually my mother would can around 300 quarts. Of course we timed our visits to Floyd County so we could bring back beans from Grandma Barnett.

In my mother's family there were nine children, and my grandfather worked in the coal mines. They were continually moving since my grandfather had to relocate, to find work in the mines. They didn't save any seeds, and it was difficult to find a place to rent with nine children. My grandfather finally moved his family to Michigan in 1950.

They moved to the muck farms as tenants. There was quite a population on the farms, with others from Appalachia along with migrant Mexican workers. The farms mainly grew onions, potatoes, mint, and sugar beets. The base pay was around thirty-five cents an hour, but the entire family was able to work at least part-time. Living and working on the farms provided them with all the potatoes and onions they could eat, something they seldom had in Kentucky. In addition, my grandfather had a full-time factory job.

In my father's family there were eleven children. My grandparents also raised an orphaned boy whose father had been killed in the mines. Later, as a teenager, the boy went to work in the same mine and was also killed. After my grandmother was widowed, she raised three of her grandchildren. So she raised a total of fifteen children.

Grandma Ella Barnett

My father's mother was the gardener and seed saver. She saved her beans in jars with a mothball or hot peppers, or sometimes she would just tie them up in an old rag. She managed to milk one and sometimes two cows a day, raise her chickens, and raise a hog or two. And she had her dogs, Coaly and Bear. Perhaps all her garden patches totaled an acre or so, but it was hard to determine on some of those hillsides.

I remember Grandma splitting wood with a single-bit axe and firing up that big black pot in the backyard, which backed up to the hillside. It seemed odd to me that lard, wood-ash lye, and water would make soap. Of course she had

Grandma Ella Barnett in the early spring of 1970, preparing to pull some sweet potato slips, cabbage, and tomato plants for a neighbor. (Frank Barnett)

used the same pot to render out lard and make the cracklings that she would toss out on an old wooden table and we would grab and burn ourselves, eating them like candy.

Grandma would can and dry beans for shuck beans. She canned her beans outside in a washtub. I had to call my mother, who is eighty-three years old, for these details. My mother said most people, before my time, would usually have two washtubs going at once to can beans. Wood, usually logs, had to be gathered; water had to be drawn from the dug well; and the tub had to be placed over a pit supported on rocks. Pieces of flat wood had to be placed inside, on the bottom of the tub, to prevent the bottoms of the jars from touching the tub, overheating, and exploding. Rags or pieces of cardboard were placed between each jar to prevent them from touching and exploding. More rags were placed on the tops of the jars to keep as much water in the tub as possible. Water was also boiled inside the house.

I remember the wood-burning cookstove Grandma had in the house and the other one she had out in the yard beside the smokehouse. There were nails, probably 20d's, driven along the top of the wall near the ceiling behind the stove with strings of beans hanging—so many you didn't see much wall between those beans.

How anyone could manage to cook or bake with a wood-burning stove is beyond me. But when Grandma was cooking with her inside woodstove and trying to dry her beans, we got to eat out on the front porch because it was so hot in the house. Green beans, green onions, greasy cornbread, fried potatoes, raw milk, and homemade butter were the norm, along with tomatoes, corn, and greens when in season. Those beans were cooked with some kind of fat pork, and for good measure a scoop or two of pure lard would top them off. And, by the way, I never weighed over 140 pounds.

I've heard lots of people say that beans dried in this manner are so much better than those dried outside. Seems I remember the kitchen being smoky. And of course, with pork of some type being fried every morning, there might have been some grease in the air. Perhaps that affected the taste of those shuck beans.

Grandma never bought a seed or a tomato plant until she was maybe eighty years old. She had always raised plants using her own saved seed. Unfortunately, we let those old bean varieties get away because we took them for granted and thought they would always be around.

Later in life I discovered some really bad beans. Nearly thirty years ago, I accepted a job transfer to the Mid-

Hudson Valley in New York. The company cafeteria served green beans, which I considered to be raw. And I missed my Martha White Cornmeal mix, but I would load up on it during my trips back home.

After two and a half years I got transferred to central Indiana, where I met an elderly gardener who was originally from Rockcastle County, Kentucky. He raised beans and raspberries to sell. He had given up raising any bean with a string years before, because no one would buy them. But he had a lot of stories about Rockcastle County and the times and beans that had been forgotten.

In 1987 I finally got to return to Kentucky. My prime consideration in buying a house was that I wanted space for a large garden with great soil, which I found in Scott County. I started to raise my grandmother's surviving bean variety after she passed away in 1990 at age ninety-four. I decided to try a more vegetarian diet to help me cope with the stress of my job.

At first, I didn't have much success locating any old-time beans, since my grandmother was the last old-time gardener on the creek. So I started visiting the local farmers' markets, with some success. But I made a lot of progress after my retirement in 2007. Two years later, I met a retired schoolteacher in Breathitt County who had driven trucks all over eastern Kentucky every summer, and he gave me some good advice as to where to travel and places to avoid (where he had never seen a garden).

Frank's Tomato Collection

Growing up, I remember my mother raising tomatoes, mainly to can in quart and half-gallon Mason jars. Rutgers, Marglobe, and Big Boy plants were readily available in our area. It seems to me there was a lot of waste involved in canning those tomatoes, especially the Rutgers, which had large white cores that had to be cut out, and the skins were hard to pull off the surfaces that had lobes.

Grandma Barnett raised her own plants from seeds she saved every year. One particular tomato stood out among the rest: a large purple tomato that had "potato leaves," a German type. I don't have a clue as to how she originally acquired the variety, but I remember it from the early 1960s. That was the primary tomato served at meals.

I remember her giving me some seeds back in 1982. They had been saved on a small piece of white cloth she had placed on top of some beans stored in a small jar in the freezer. I took the seeds with me to New York State and was successful in growing the tomato. However, after several moves, I lost the seeds. I later tried to obtain more seeds from my grandmother, but it was too late; they were all gone.

A lot of people have told me that they are looking for the tomatoes their parents or grandparents raised years ago. The only way they will be successful is to locate someone else in the area who got the tomato from their relatives and used the same selection process for saving seeds over the years. But it is unlikely they will ever find exactly the same variety.

I accidentally found Coffey's Greenhouse in Russell Springs back in 1991, while on a trip to Lake Cumberland. They were selling Jack Miller tomato plants, a local variety that is a large, pink oxheart type. They also had a wide selection of other tomato plants, including Mountain Spring, a white variety; a Black Russian variety; and the Cherokee Purple.

During my search for heirloom beans, I discovered that nearly all the people I met who were growing them were also raising heirloom tomatoes. I met Bill Best, and he mentioned Vinson Watts, so I went to Morehead to meet Vinson. Over the next two years, we spent hours talking and became good friends. He sold his tomato plants from the side porch of his house, and I initially bought four plants.

I had traded heirloom beans with Tim Amburgey from Ball Fork in Knott County. He was also growing a tomato from John B. "Quincy" Adams of Carr Creek/Pinetop, a retired miner who had recently passed away. This was a prized tomato for me, since it is the closest to my grandmother's that I have ever grown.

I bought some beans from Ralph Hollin in Manchester, Clay County, through an advertisement in *Kentucky Explorer Magazine,* and we later traded some other bean varieties. He gave me two tomato varieties that his relatives, the Maxie family, had grown for as long as he could remember—and he was seventy-three. After World War II ended in 1945, some family members migrated to Indiana to find work. Of course, they took their family tomato varieties with them, and years later they brought them back to Clay County, where they had been lost. Uncle Owen (Maxie) and (Cousin) Virgil "Buck" Maxie are both excellent varieties that I have grown for the past two years.

After reading an article about seed savers that appeared in the *Ashland Daily Independent* newspaper on October 3, 2012, I called one of them, Kevin Scraggs of Greenup, and went to visit him and his wife, Cathy, and traded for his five local tomato varieties: Carrie Claxon (recently accepted nationally), Elliott County Huge Orange, Elliott County Huge Pink, Striped Potato Top, and Larry Easton Family

Pink. As mentioned in the article, Kevin has sixteen varieties of tomatoes and beans with local roots. In 2013 I raised five of his bean varieties and the two tomato varieties from Elliott County.

Someone at the 2013 seed swap in Pikeville gave me the phone number of John Rawlins, who lives near John's Creek. He had a bean variety that my grandmother had raised—the Striped Double Hull White Cornfield bean. So I gave John a call and stopped by during a trip back to Floyd County to visit relatives. Like many heirloom bean growers and seed savers, he also grows a large yellow tomato his mother got from a relative back in 1965 and continued to raise until she passed away. I planted the tomato in 2013 and thought it was one of the best yellow tomatoes I have ever grown.

In Harlan County at Cranks I met Roy Farmer, who raises a huge garden every year. We traded beans and tomato seeds of the variety he calls the "Pure Free Man," which I raised in 2013. It is a red tomato the size and shape of a softball—an excellent tomato.

Altogether, I have found more than twenty heirloom tomato varieties that have been grown in Kentucky for decades. I mainly limit myself to growing Kentucky heirloom tomatoes from gardeners I have met. In most cases, I have seen the mature, ripe tomatoes in their gardens. I have found that the growers who give so much attention to saving their bean varieties take the same care when selecting and saving their best tomatoes for seed.

Recently, one of my high school classmates sent me a new bean variety from Buncombe County, North Carolina, and a new Candy Roaster squash from Upper Crabtree, my home commu-

Candy Roasters ready to be split open for seed saving and/or eating. The seed came from a champion grower in North Carolina. When I was growing up, we grew pumpkins for cattle feed and jack-o'-lanterns, but we grew Candy Roaster squash for pies. Many think these pies are far superior to pumpkin pies.

nity there. One of the reasons many of us have continued to save seeds over the years: the fascinating stories behind each one of these heirloom varieties. Frank says that many of the heirloom bean stories he has heard are not as involved as the following one about an Australian heirloom tomato that became an Indiana tomato before finally becoming a Kentucky heirloom. Each variety has its own story. And with some, it is simply a matter of stealing from someone else's garden.

The Jack Miller Tomato

In 1995 I discovered the Jack Miller Australian tomato at Coffey's Greenhouse on Route 127 in Russell Springs. It is a large, pinkish red oxheart type that is very sweet. The plants must be caged, since they are very indeterminate. I have saved the seeds and raised a few plants every year, but I decided I wanted to learn about their background. So in the spring of 2009 I drove to Russell Springs and went back to the greenhouse, which is now called Anna's Garden Center. Anna told me that the Jack Miller was the most popular tomato plant in the area, outselling all her other tomato varieties. And Jack Miller was actually a local person who raised his tomatoes every year and supplied Anna with the seed to grow her greenhouse plants. According to Anna, everyone in the area knew Jack and thought a great deal of him, and she gave me his phone number.

When I returned home, I did an Internet search to find Jack's address and discovered that he was an attorney. I gave Jack a call and told him that I had been raising his tomato for years and wanted to learn its background. I also mentioned that I was trying to collect heirloom beans. He said he had plenty of bean seed in the freezer that he would be glad to share and invited me for a visit.

The first nice day I drove down to see Jack. He had given me good directions, and his place was easy to spot, since he had told me to look for a five-acre lake with more than 140 Canada geese in his front yard.

Well, everyone knew Jack because he had retired as the district judge at seventy-five years old; he was now eighty-eight. Jack had started college at the University of Kentucky, but after World War II broke out, he quit college and joined the navy. After the war he went back to UK on the GI bill ($100 a month) and graduated from the law school in 1950. He returned to Russell Springs to practice law, later became the county attorney, and retired after serving seven years as the district judge.

The tomato had come from Jack's future brother-in-law,

who had majored in agriculture at Indiana University and became the county extension agent in Green County, Kentucky. One of his best friends at IU, perhaps a fellow agriculture major, had taken a trip to Australia and eaten some tomatoes he really liked, so he brought home some seeds. He gave the seeds to his mother in Indiana, and she grew the tomatoes and gave them to her son, who gave them to his friend (Jack's future brother-in-law), who gave them to Jack. Out of nearly a peck of tomatoes, Jack saved two of them for seed. From those two tomatoes he got only twenty-four seeds. The next spring Jack asked his neighbor who had a small greenhouse to raise the seed. They shared the yield of fourteen plants. Over the next few years, Jack continued to pick his choice tomatoes for seed and had his neighbor raise the plants.

Perhaps because of the popularity of the tomatoes, Jack took his seeds to Coffey's Greenhouse in Russell Springs to raise. When the greenhouse was sold to Anna, Jack made the same arrangement.

I never got a definite answer as to how many years Jack had been raising his tomato. However, I am at the age when I think something that happened twenty years ago actually took place forty years ago. The bean seeds from Jack's freezer had dates ranging from 1983 to 1987. I got a total of nine bean types, two of which were lima beans. It is always interesting to get a story along with an heirloom bean. Jack had an Aunt Sal's Purple Hull pole bean, which turns green when cooked. Also, he had a bean from the local preacher who had traveled to Oneida, Tennessee, to preach at a revival. He stayed with a family there and was so impressed with the beans they fed him that they gave him seed to bring back to Russell County.

Judge Jack Miller of Russell Springs raised the Jack Miller Australian tomato for years, a large pink-red oxheart type. (Frank Barnett)

I thanked Jack for a good visit and all the seed and promised that I would try to grow each bean out and share the seed with other heirloom seed savers and that I would continue to raise his tomato.

Judge Jack Miller passed away on March 9, 2011, and his funeral was held at the Fairview Separate Baptist Church on the twelfth of March.

Trip Notes

Frank Barnett has accumulated an impressive collection of heirloom beans, primarily from eastern Kentucky. Since his retirement, he has logged thousands of miles in search of heirloom beans and their growers. In Breathitt County alone he has found more than fifty varieties and is still looking. His trip notes pro-

111

vide a lot of insight into the history, culture, and folkways of eastern Kentucky.

When he first began collecting, Frank added eighty-seven bean varieties to those he already had from his family, largely from six counties in eastern Kentucky—Breathitt, Floyd, Knott, Laurel, Morgan, and Wolfe. He is the only active collector at this time whose entire life is devoted to making contacts and following up on them. He meets the growers personally and trades seeds with each person he collects from. He is constantly visiting his contacts, especially those in eastern Kentucky and even in West Virginia and Virginia. His old 1999 Toyota Camry has around 350,000 miles on it and is still going strong. While most other collectors pride themselves on having a few dozen varieties, if that, Frank's collection numbers in the hundreds.

Every bean has a story, as his many tales attest. He's good at recording names and the stories of how he collected each bean. We often talk about how different beans might have come by their names. Frank recently told us about the naming of the Josara Fall bean he had found in Harlan County near Wallins Creek. Frank thought the name was interesting and assumed that, like most beans, it had been named for a woman, the gardener in the family. Frank said he thought it might have been an ethnic name because the mines had drawn workers from all over Europe. But the man who gave him the bean told him it had come from a coal camp where a man and his wife had kept a garden and raised the bean. He couldn't recall their last name, but their first names were Joe and Sara. So Joe and Sara's fall bean became the Josara Fall bean.

Many beans are named after a gardener. Frank thinks it's because the gardener recognized an odd, different bean—a sport or a mutant—saved the seed, and then developed that variety. Frank has three beans called Aunt Maggie, all of them different beans from different areas and different Aunt Maggies. The same

The Josara Fall bean, a brilliant speckled red and white bean acquired by Frank Barnett in Harlan County near Wallins Creek. (Frank Barnett)

with Big John beans. He has five varieties of them. He says most people have an Aunt Maggie or an Uncle John in their families, but Frank has them on both sides!

Here's a sample of Frank's trip notes from November 2010:

Finally, deer season has given me a good reason to stay away from Robertson County and write about my trips of the spring and summer.

I had decided to take a day trip to Harlan County on the first nice spring day. I had done my research on the Internet and made a list of places to check for garden seeds.

Every place on my list was a strikeout. The farm store in Harlan had gone out of business, along with Maynard's in Cumberland. Only one store in each town carried garden seeds from the Knoxville Seed Company. All other stops in Harlan no longer sold garden seeds. However, the manager at the last hardware store in Harlan suggested that I should check out Matt's Feed and Seed store five miles out of town on Highway 38 at Ages.

Matt's Feed and Seed was a small block building, and they had eight heirloom bean varieties for sale—the 1001 bunch, Arnold Williams, Aunt Beth, Josara Fall, and others. The beans were in pint baggies along with a mothball. Matt Napier said Dewey Harris in Laurel County near East Bernstadt had raised the beans, and he gave me Dewey's phone number.

Dewey has a green and white 1967 Dodge pickup truck that he bought used in 1970, and he has been driving it for the past forty years. The truck is well known by customers to the London Farmers' Market. Besides being a gardener, he had farmed tobacco and laid brick/block for over thirty years before his retirement.

Dewey started collecting beans twenty-five years ago. One of his first beans was the Aunt Beth, which was raised by his great-aunt for sixty-eight years until her death at age ninety-four and then raised after that by his aunt. He has raised it for the past twenty-five years. He gave me seeds of twelve additional beans. I went over the beans I had to trade and asked Dewey which ones he would like to try. I mailed those to him the day after I got home.

Dewey mentioned that he does a lot of shuck beans. They break and string the beans and lay them on old sheets or old blankets on a flatbed farm wagon to dry. After drying, they used to bag the beans and place them in the freezer, but in the past few years they started just heating the dried beans in the oven until they are hot to the back of the hand, letting them cool, bagging them, and leaving them in the kitchen.

Due to the weather, I didn't get to raise all the varieties I had gotten from Dewey. I had only a few seeds of some varieties, and I didn't want to risk them to the bad weather. However, I was well pleased with the varieties I did raise: the white cornfield beans—Aunt Beth, Arnold Williams, and Shag Howard—and the two fall beans—the Kentucky Mountain (a small, striped fall bean from Harlan) and the Rocky Knob (a large, beige cut-short from western Virginia).

Dewey has worked hard his entire life. He plants about five acres of vegetables and hoes most of it out by hand. He uses no plastic ground cover or drip tapes, so this hot, dry summer hit his garden hard.

My next big trip was during the July Fourth weekend to Breathitt County for the Route 52 yard sale, which stretches from Jackson to Irvine, a distance of thirty-seven miles. There were a few beans for sale, but they were all commercial White Half-Runners. However, someone gave me the name of a gardener on Route 30 that I should visit.

I stopped at the house near the Old Regular Baptist Church at Shoulderblade and met Paul Griffith. Paul is a brother-in-law of J. B. Mullins, whom I met earlier this year. I have raised several of J. B.'s beans for the past two years, and they have all been good tender beans. Paul worked in the Michigan steel mill along with J. B. before they both moved back to Breathitt County. Paul asked for my address and said he would mail me some seed later, when he felt better. Besides, they were in the process of cleaning the freezer after the compressor had quit. I offered him some money for postage, but he refused. I thoroughly enjoyed my visit with Paul.

A big surprise came nearly two months later when Paul's brother, Granville, called me at home. Paul had been in the hospital, so he had charged Granville with the task of getting some seed to me. I told him I would mail him a check for the postage, but he offered to deliver the seed personally on his way back to northern Kentucky. Granville arrived with an entire freezer basket full of more than a dozen jars and bags of seed corn. Only six beans were labeled, and none of the bags of field corn. I believed two of the bean varieties were commercial, but I thought the one labeled the Old Fashioned Bill Johnson might be an heirloom. I thanked Granville and mentioned that I would be going to Breathitt next month to visit Arthur Johnson, who would be making sorghum, and said I would stop by to see Paul and thank him for all the seed.

So in September I went back to Breathitt to visit Arthur Johnson on Route 30. I had met him at the Quicksand Farmers' Market earlier this year and bought a seeded watermelon from him—the best watermelon I had the entire summer. He had also given me some yellowed pods of a long, white cornfield bean from Long's Creek raised by Jim Deaton, who is in his mid-eighties. A local paper had recently printed a photo of Jim carrying out two buckets of beans from his garden. Of course, I plan to raise the bean next year.

Arthur had saved back a quart of sorghum for me; the past week he had run off forty-nine gallons. He also had a bag of pink six-week bean seeds for me, a bean that he and his family have raised for nearly fifty years. When he found out that I hadn't been able to plant any greens this year due to the hot, dry weather, he cut three bags of three different greens for me. Also, his wife gave me some perennial flower seeds. They invited me to come back and visit again.

Bill Best with a handful of leather britches. (Dobree Adams)

Carbonized archaeological specimens of the common bean (*Phaseolus vulgaris*) and their uncarbonized heirloom counterparts arranged by Bill Best: top row, regular beans; second row, cut-short beans; third row, fall or October beans; bottom row, grits beans. A Kentucky Wonder bean is shown in the top row, extreme right. Burning preserves the archaeological specimens, but it shrinks them in size and chars away color. (Hayward Wilkirson, University of Kentucky)

Frank Barnett Cut-Short bean. (Michael Best)

Greasy beans on my favorite rock on our back porch.

A medley of heirloom beans.

Rebecca Pridemore, a full Cherokee and great-grandmother of Frank
Barnett's Grandma Ella Barnett. (Collection of the Barnett Family)

A basket of Frank Barnett Cut-Short beans, freshly picked from the garden. (Frank Barnett)

Vinson Watts tomatoes growing in a cluster.

The Willard Wynn Yellow German tomato, one of our heirloom favorites.
(Dobree Adams)

A meal fit for a king!

Four generations at the Lexington Farmers' Market: Bill Best, son Michael Best, grandson Brian Best, and great-grandchildren Greta, Peter, and Lilah Hess. (Christina Best Hess)

Heirloom beans and tomatoes growing in our field. (Dobree Adams)

The large red Zeke Dishman tomato. (Dobree Adams)

Boyd Smith Yellow German tomatoes ready for the market. (Dobree Adams)

Ira Wallace of the Southern Exposure Seed Exchange at a seed-swapping event in 2011.

Susana Lein's Kentucky Rainbow corn, selected and bred from dent corn seed gifted by Daymon Morgan and grown for generations by his family in Leslie County. (Phillip Gilbert, Disputanta, KY)

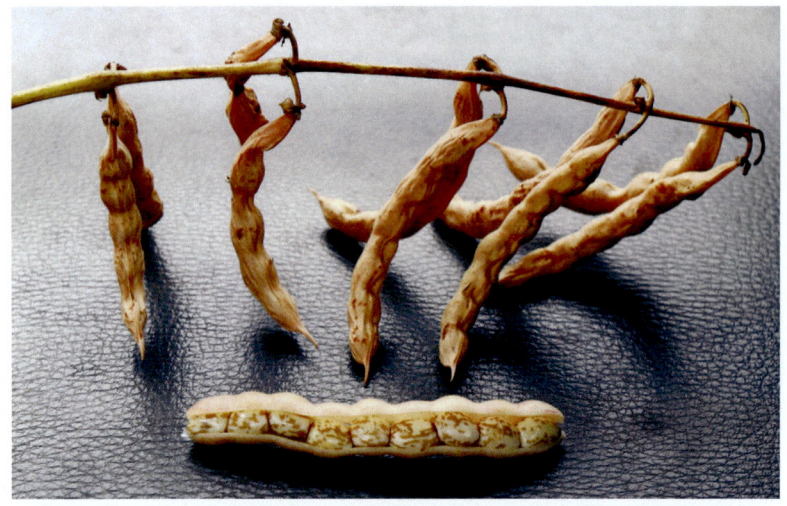

"Soldier beans," so named because the pods line up in formation. In the foreground is a "bust-out" bean. Cut-short beans sometimes break apart or bust out, scattering their seeds, if they dry out and then become wet again. The "bust out" is another way to identify the cut-short.

Kentucky Flat Tan Field pumpkins, an old Kentucky heirloom found in Lincoln County, 2015. Considered one of the best culinary pumpkins in the world, this variety probably originated in the mid-nineteenth century. (Dobree Adams)

Virginia Gourdseed corn, originally from the Virginia–North Carolina border, was used by early settlers to breed new varieties. This specimen was grown at Fort Boonesborough State Park with seed from the Southern Exposure Seed Exchange. (Barbara Grant Elliott)

Passing it on: Greta Hess, five years old, learning to string beans.

Before leaving Breathitt, I stopped by to see Paul and to thank him for all the seed he had sent me via his brother. I promised I would plant everything but noted it might take me a few years. I told him I was surprised by the amount of seed he had sent me, and he replied that he had given me his word. Paul had long forgotten who had given him most of his beans over the years; nor could he recall the names of the beans. However, he did remember the year his parents and all the children raised and sold over 100 bushels of green beans at $4 a bushel, which bought clothes and shoes for the children to return to school.

A few weeks later I made another trip back to Breathitt County after receiving a call from Judy Thorpe of War Creek near Van Cleave. Last spring she had given me two beans that her grandmother, Mertie Hollon, had raised until her passing in the 1960s in her mid-eighties: a brown speckled fall bean and her own white half-runner. I had raised both these beans this summer and liked them both. Judy said she had raised two additional good beans this summer and would share some seed with me: a white crease-back from the Shelton family and a white ram-horn type from Turner's Creek. I took her seed from two white cornfield beans I had raised this summer: the Granny and the Square House.

Going on to Floyd County, I dropped by to visit Peco Hall, who had given me four bean varieties earlier this year, three of which I had raised this summer: the Spurlock white cornfield, a white greasy cut-short from Menifee County, and a small speckled greasy bean from his late neighbor John Morgan Stumbo.

It is interesting that Peco raises his beans on commercial hybrid field corn. After he is finished picking beans, he gives the corn away to neighbors and relatives who have livestock to feed. He was raising a new bean given to him last year from the family of Grandma Hagan of Turkey Creek at Langley in Floyd County, who had passed away the prior year at the age of ninety-two. She had left a pint jar of seeds of her white cornfield beans in the freezer. The family shared these with Peco because they knew he would plant them. Peco picked several of the yellowed pods for me and also gave me some dried seed. This is another must-grow bean for next year.

Peco is eighty-three years old, and he shared a story about his neighbor John Morgan Stumbo, who also raised his own heirloom pepper plants. He had given Peco a few plants one year, and the next year Peco asked John for some more of those good peppers. "You didn't save any seed?" John asked. "Well then I reckon you'll be doing without any peppers this year." Peco said John had taught him a good lesson: always to save his own seed. John wasn't a man you wanted to go back and ask again for the same seed.

At the Morgan County Sorghum Festival at the end of September, I bought three additional bean seeds from Wanda Hamilton—the Marfeds, a large white bean her uncle had raised; a large white greasy cut-short; and a brown greasy—all of which have been raised by her family in Morgan County. I grew two of Wanda's beans this past summer that I had bought last year, and they were both outstanding—the Square House, a long white cornfield bean that grows in bunches, and a striped cornfield. She and her daughter sell produce at the local farmers' market on Thursdays.

I have been to Powell, Lee, Knott, Floyd, Perry, and Harlan Counties this summer, and in all these places I found beans for sale from the Western North Carolina Farmers' Market at Asheville. And in Harlan County, beans were coming from Grainger County, Tennessee. On a trip to Whitesburg, I found that the Golden Apple Fruit Market, south of Whitesburg on Route 15 toward Jenkins, also buys its beans from the market at Asheville. They go to North Carolina twice a week to buy produce.

I came back through Hazard and stopped at the farmers' market there. Three vendors were selling beans raised by the Mennonites of Casey County.

I made another stop at Smitty's Market on Combs Road in Hazard. He had several green beans out of the Western North Carolina Farmers' Market. I found a long speckled greasy that appeared to be different from the others I have, and he gave me five yellowing pods. He has cut back on trips to North Carolina and no longer grows any of his own beans for sale because he can't hire the help he needs. After operating his market for thirty-two years, he doesn't buy a lot of locally grown beans for resale anymore because most customers want commercial-looking beans with no spots, bends, etc.

I have met many old-time gardeners, and they have two things in common: they are all willing to share their seeds to "give me a start," and they all want their seeds to go on after they leave this world. I always mention that I am a grower, an eater, and a seed sharer. I don't use the word "collector," which would imply that I keep what I find to myself. I'm willing to share any seed I have to "give a start" to someone who is also a grower, eater, and sharer.

7

Seed-Saving Legacies

While doing research for this book, which involved a lot of traveling and talking with different people, I constantly ran across individuals who had interesting stories to tell about their lifelong gardening and seed-saving habits. Thanks to the Internet, many others were able to share their experiences via e-mail. Some of these stories were so significant that I asked the individuals to write them down in their own words. There is no way I could capture what they had to say and provide full appreciation for the context of their lives.

For some, their stories revolved around lifelong habits of thrift, respect for family customs, and a desire to hold on to cultural traditions, however far they might have moved from their birth homes. For others, it was about renewing traditional ways that had temporarily been abandoned due to relocation for a job, to take care of an ill family member, or some other situation beyond their control.

Bill Leach

I have known Bill Leach since I was a freshman at Berea College. A few years after that, when we were both serving as trustees of Pine Mountain Settlement School, we started trading heirloom

John Coykendall, a renowned seed saver and master gardener at the
Blackberry Farm resort in Tennessee for many years. He grows endangered
beans one or two seeds at a time to maximize the chances of success. John has
brought some seeds back from the brink of extinction, including the Noble
bean described on pages 78–81.

bean seeds. We were simply adhering to a long-held tradition in
both of our families. Today I am still growing many of his seeds
and sharing them around the world on the Sustainable Agriculture Center website.

Dr. William Leach, now retired from his position with the

US Public Health Service, continues his gardening and seed saving, passing on a multigeneration seed-saving tradition to his children and grandchildren that has been unbroken since the Civil War.

"Tradition! Tradition!" Thus starts the Broadway musical *Fiddler on the Roof.* I was born into tradition and spent the summers of my youth learning about it. Most of us are born into a tradition, perhaps a family tradition, as I was. Mine went back three generations before me, to my great-grandparents. My great-grandfather William Creech obtained a land grant, through his service in the Union army during the War between the States, in an isolated valley in southeastern Kentucky. By the time his family had grown by two generations, he was anxious to improve educational opportunities in his valley, and he eventually donated the land on which the Pine Mountain Settlement School came into being in 1913.

His wife Sally was a strong believer in learning by doing and in doing what one learned. Late in her life she undertook to weave coverlets for each of her children. She used her loom made of poles. She wove with the flax she grew, retted, and spun; with the wool she sheared from the sheep, cleaned, carded, spun, and dyed with materials she had grown or gathered. Some people still living in the valley practice the arts and crafts associated with the creation of fabrics. Sally also gardened—a necessity in her day. When William and Sally moved into the valley, she brought seeds from her parents' fields. Perhaps some of the seeds came with the first settlers in Kentucky; perhaps some were obtained from the Cherokee residents already in the area. When her children moved to independent living to start their own families, she furnished them with seeds from her garden. The site for the cabin William built for his family was selected because of the availability of fresh water, bottomland suitable for agriculture, and the presence of a canebrake, a native Kentucky version of bamboo that had many uses in pioneer living.

I became aware that my mother planted seeds from Pine Mountain: white corn for cornmeal, yellow corn for livestock, speckled fall beans, and a white bean variety that was popularly known by two names—white drying or cornfield. I learned about the history of the seeds; I experienced some of the hardships connected with growing them. Hoeing corn in which beans had also been planted was a real chore for me: trying not to chop off the bean runners that had not climbed the cornstalks, stopping to try to wind them onto a stalk, for I knew that picking beans from runners on the ground was both messy and possibly dangerous (I thought). I had a fear of the rattlesnakes and copperheads that might be lurking in the tangled vines. That fear was not totally unfounded; I'd come face-to-face with copperhead snakes among stores of dried onions and under sheds, and once I almost stepped on a rattlesnake as I fetched milk cows from the mountainside pasture.

Cleaning up the brush and sticks used to support vining beans and peas was not a pleasant chore either. One rainy day I was told to remove a row of brush and sticks— the plants they had been supporting had dried up—and cut them up so they could be burned in the cookstove. The one promising aspect of this assignment was that I had not been told how to do the cutting, so I elected to use my grandfather's double-bit ax. The ax tended to bounce off the wood, so I used a foot to hold the sticks in place. Well, the ax slipped, and my foot was in the way. I managed to split through my new summer boots and my small toe. I tore a handkerchief into strips and bound up the toe to stop the bleeding. For several days wearing the boot was very uncomfortable, but I did not get an infection in the toe, and no one commented about the cut on the boot. The pair of boots is long gone; the scar is still with me.

My experiences associated with my work on my grandparents' farm formed a collection of very rich memories I still cherish.

My mother enjoyed gardening and raised a variety of vegetables that included the seeds from her parents as well as from other people we knew. I helped with her garden and learned the source of several of the seeds she saved and planted year after year. I contributed a couple of seed varieties that I thought were interesting. Among the seeds added to those from Pine Mountain were what we called the Lee bean, a brown-colored snap pole bean that was a family heirloom bean of a lady we knew in Monticello, Kentucky. We located two speckled lima bean seeds at a local store, also in Monticello, the larger of which is now sold as the Christmas variety lima bean. Someone gave my mother a bush bean seed that produced within six weeks after planting. Mother called it Ant-Egg because of the shape and color of the seed.

After completing my education, I moved with my wife and children to Washington, DC, to work as a member of the US Public Health Service. I brought some family seeds with me, and part of the backyard of our suburban house became a garden. For forty-four years, most of those seeds have been planted in my small garden plot. During the early years, the yield was only enough to save seeds for planting the next year; now produce is shared with neighbors and preserved for use during the winter months. Some seeds from other relatives have been added to my seed collection; other varieties have been lost—the yellow corn, a bean known as Lizard Toe, the speckled tender-hull fall bean. Space is insufficient to plant each of the varieties I have, so I rotate what gets planted each year. Since few people garden in our neighborhood, where many are retired, the demands of birds, squirrels, and chipmunks on *my* garden now make planting corn a high-risk venture. Rabbits enforce the need to replant bean seeds—seeing short stubs only a half inch above the ground where three or four young seedlings proudly stood the day before can be disheartening.

For several years since I retired, my older son has shown up each spring to plow my garden plot. His sister and brother often pitch in to trim bushes and clean the yard and tool shed, where I keep my grandfather's double-bit ax and a mowing blade I used to mow alfalfa at Pine Mountain years ago, along with the tools that are useful for tending small gardens and flower beds. This spring we dined on white cornfield beans and made soup with speckled lima beans. This year, my son—the one who knows how my grandmother and mother prepared fried corn—took some white corn seed to plant in his garden near Detroit. I shared with him a peculiarity of that corn: cobs of genetically unmixed ears are white, and cobs of mixed ears are red, so don't save kernels from red cobs for seed.

Tradition—yes, indeed!

Gary Millwood

A longtime seed saver who worked diligently to discover, grow, and distribute Kentucky heirloom tomatoes, Gary Millwood shared his findings with Maria Stenger, who has been growing Kentucky heirloom tomatoes since 1997 and selling them through Blue Ribbon Tomatoes on eBay.

Gary's wife, Katharine, contributed these thoughts after Gary died:

> Gary's interest in gardening began as a child, spending time during the summer with his grandparents. Speaking of his grandfather, he wrote, "I never realized how much influence his love of the land and his knowledge would follow me to adulthood."
>
> Due to serious health issues, Gary had to retire from Bellewood Presbyterian Home for Children near Louisville in 2000, having served in various positions at Bellewood for twenty-five years. He renewed his interest in gardening "because of the therapeutic value" and devoted his time to researching, growing, and sharing Kentucky heirloom tomato seeds with folks all over the world. He discovered many Kentucky tomatoes that are now listed in several seed catalogs. He loved the stories that came with each discovery. (One definition of an heirloom tomato is a tomato with a story attached.) He was most proud of the story that goes with the discovery of the tomato Aunt Lou's Underground Railroad. After extensive research, he was able to connect it with the Kentucky-Ohio Underground Railroad—thus the name.
>
> He was known as Gary to some, Pop-Pop or Papa to his grandchildren, Briarpatch to his fellow clowns and the Make-a-Wish families, and Tomato Man to his many on-line tomato friends. He died on May 29, 2013, just four days following our fiftieth wedding anniversary.

Gary contributed the following before his death:

I was born at home in Valley Falls, a textile mill community, in Spartanburg County, South Carolina, on September 29, 1936. When I was four we moved to Chesnee, South Carolina, where my father was a supervisor in a textile mill. Three years later (1943) we moved to Beaumont, another mill community in Spartanburg. Beaumont's mill was in full production, making heavy duck cloth or duck canvas, commonly called "canvas" outside of the textile areas. The material was being used in the war effort.

I remember when President Roosevelt died in Warm Springs, Georgia, and the train carrying him back to Washington passed four blocks from our house. I stood on the bank, as did hundreds of folk, and watched the Southern Railway car pass, the doors open and his casket in view surrounded by soldiers from all the armed forces. I remember it as a sobering moment even as a youth.

My grandfather, Edward Paris Cartee, was a tenant farmer and always lived in the Cowpens area of Spartanburg County. The Battle of Cowpens was a turning point in the American Revolution and was a very short distance from one of the farms he lived on. Grandpa had a limited education, but he loved the land and cared for it with a passion. He practiced "green" before anyone ever thought of it! And you could see the results in the crops of his fields.

They had a cow (treated like a member of the family), pigs, chickens for eggs and meat, and two mules. Grandma Essie Lands Cartee cooked on a woodstove and canned meats, vegetables, and fruit. She also dried apples and peaches. They had a well and a springhouse where they kept butter, eggs, and other perishables. They had a raised mound area out back of the house where they had heads of cabbage and other greens covered with straw, a tarp, and soil, which kept them from freezing. Apples and sweet potatoes were in another building. Smoked meat and hams were located in the smokehouse.

I don't remember ever seeing a tractor in those days, unless it was at a peach orchard, and there were plenty in those days. The county produced more peaches than the whole state of Georgia; that's not the case these days.

I became acquainted with friend Maria Stenger about four years ago. With the heart and health issues I have had over recent years, I wanted to have someone who was dedicated and committed to preserving the Kentucky heirloom tomato varieties. She was touched when I told her that I wanted to share everything I have with her. Merlyn Niedens was doing his part by growing and listing them with seed companies, and that was good. [Two seed companies—Baker Creek of Manfield, Missouri, and Southern Exposure Seed Exchange of Mineral, Virginia—listed twenty-five to thirty varieties Gary had shared with their grower, Merlyn Niedens, Gary's mentor and friend, who died in May 2010, working in the garden with his tomato plants.] So I have shared most all the varieties, old and new ones, that have come my way. Maria has discovered a number of varieties as well, and we hope to discover more. There is nothing more exciting than discovering a tomato that has a history—a time and place and real gardeners who nurtured and preserved it so that others could benefit from their efforts.

According to Gary, his interest in growing old tomato varieties lifted his depression following heart surgery and retirement. In its place grew a passion for raising tomatoes and sharing seeds with folks around the world. Gary passed along the following article he wrote about his interest in growing heirloom tomatoes:

Growing up in town as a child, I had limited exposure to growing things other than a few flower plants my mother grew and some purchased tomato seedlings my father grew. As children we helped with the watering and care of the plants.

In days gone by, growing vegetables and seed saving became a tradition, or better still, a necessity. These seeds provided hope and prosperity for farm families. My grandpa, Edgar Cartee, was a tenant farmer in South Carolina, and in my early years I spent some weeks in the summer visiting and enjoying the country life and helping as I was capable. One thing for sure is that I was soaking in all those summertime experiences. I owe so much to him and my grandma, Essie, for having the patience to tolerate a growing city boy who always had so many questions.

My life's work has been devoted to serving the church. With the exception of six years and a few months working as director of education for a church in Greenville, South Carolina, I have served in an administrative capacity in three children's homes in three states.

In 1978 I returned to the Bellewood Presbyterian Home for Children in Anchorage, Kentucky, where I had worked some years before. Aside from the responsibilities of supervising staff and children, I was responsible for the farm, twenty-two head of beef cattle, pastures, hay fields, garden, buildings, and grounds.

The garden served as a major source for fresh vegetables for summer meals. This was the time I began using methods I had observed from my grandpa. Like all gardeners, I had to deal with the seasons, weather, some trial and error, and accepting good advice when it was given. I learned the varieties that produced and did well for us. The county extension folks were very helpful to me during this time.

The importance of the farm operation transitioned with the retirement of the director and the arrival of a new director. We initiated new programs that focused more on services to children and families. We continued with a small garden to provide summer experiences for children who enjoyed the activity, and I was involved as much as possible.

When I began having serious health problems, I renewed my interest in gardening because of the therapeutic value. In the early 1990s I had to undergo angioplasty, and in 1995 I had a heart attack and a triple bypass. My recovery was remarkable, and I was back at work early. Four years later I had a second heart attack and a second triple bypass; this time, my recovery was long and difficult, and I had to retire from my work.

The winter of 2000 I decided I would limit my gardening to growing mostly heirloom tomatoes. Not too long after this, I attended the funeral of a dear friend who had grown up in eastern Kentucky. There I met Jerry Cantrell, the new director at Bellewood Home. He had heard that I was interested in growing tomatoes and mentioned that his mama Lettie, of West Liberty, had a tomato I might enjoy growing. She sent me seed, and I loved it so much that I have been growing it ever since. Lettie died in 2005 at age ninety-six. Granny Cantrell's German tomato is the only tomato she ever grew. Baker Creek and Southern Exposure Seed Exchange listed it in their 2007 catalogs.

Byron Crawford was visiting my friend Nancy Theiss at the Oldham County Historical Center, where I volunteer growing some old plant varieties, including heirloom tomatoes. This interested him, so he wrote an article for the *Louisville Courier-Journal* about my interest in old tomato varieties. Julie Maruskin of Winchester read the article and contacted me about her Coy family's Depp's Pink Firefly tomato. This tomato from the 1890s originated in the Glasgow, Kentucky, area. It is one of the most beautiful tomatoes I have ever grown, reminding me of a large Christmas ornament. Underwood Gardens lists it in their catalog now.

I am a member of several gardening sites. I correspond with several serious tomato growers, including Doug Zucknic in Romney, West Virginia. He was searching for several seed varieties, and I was able to find them for him. Last year he sent me seed for the Purple Dog Creek heirloom tomato. He acquired the seed from a gentleman who had visited Kentucky with a church group on a mission trip to repair housing for the elderly. The host church provided a covered dish dinner on their last evening there. Members brought their favorite dishes, and one member brought his wonderful Purple Dog Creek tomatoes! Seeds were shared, and Doug sent me a few last year, but they failed to germinate. I am treating my seed with care this year in hopes of growing this tomato.

One of three Ashlock brothers who served with George Washington during the Revolutionary War settled in Kentucky. Carl Ashlock, now of Franklin, North Carolina, is descended from that patriot. Carl and his father and grandfather farmed in Kentucky, where they grew a large pink tomato variety, the seed of which Carl shared with me and others. He said he hoped others in Kentucky would be interested in growing and saving Grandfather Ashlock's family heirloom.

Last summer I tasted the Unknown Kentucky Heirloom, aka the Kentucky Pink Stamper, at our CHOPTAG (Cincinnati Heirloom Open Pollinated Tomato Associate Growers) meeting. Both Earl Cadenhead and Carolyn Male shared a few seeds with me. I learned later that Mary Klacson of Eugene, Oregon, had received a tomato called the Unknown Kentucky Heirloom from James A. Stamper, who resides in Dwarf, Kentucky. He said his family had been growing this tomato and a Kentucky white pole bean for as long as he could remember. Since the tomato had no name, Mary named it the Kentucky Pink Stamper for the Stamper family.

My friend Al Anderson of Troy, Ohio, and I have been sharing our favorite heirloom tomatoes for the last several years. He and several of his friends grew better than 350 varieties in 2006. His friend John Siegel has been growing the Kentucky Wonder tomato for years and another called the Kentucky Striped. I have seeds of these two and will be growing them this season.

While visiting Berea this past winter, I met Sharon Patton, who grew up in Jennings Hollow near Monticello in Wayne County, Kentucky. She couldn't wait to tell me about her family and her grandmother. She said they had an all-time favorite tomato her grandmother grew that came from her grandmother's grandmother, who immigrated from Ireland. It has not been named, so I think we are going to call it Lizzie's Irish Eyes!

Seed saver Gary Perkins of Wayland, Kentucky, grows and shares many of his seed with SSE (Seed Savers Exchange) members. Gary has shared a number of his heirloom tomato varieties with me. I especially enjoyed the Lenny and Gracie's Yellow and the Black Mountain Pink varieties.

Black Mountain Pink: Round pink fruit one pound or more with very good flavor. It was originally collected and named by Austin Isaacs of Richmond, Kentucky. It dates to a Mr. Harrison, who discovered tomatoes growing at an abandoned homestead in 1933. It is an indeterminate, with seed available from Marianna's Heirloom Seeds.

Lenny and Gracie's Kentucky Heirloom: Large yellow ribbed fruit with a pink-red blush and a juicy, fruity flavor. From the Johnson-Magoffin County area of eastern Kentucky. Lenny and Gracie's Kentucky Heirloom was originally collected and named by Roger Wright of Hamilton, Ohio. There is a red version as well. Indeterminate.

What I think is important as I consider each individual story is that someone thought these tomato varieties had some good qualities and made attempts to save seed. Many were shared with family, friends, and others. Each time I discover a variety like one of these, it is like discovering a lost treasure. Growing and tasting them provides experiences that others have had over the years. For me, it reminds me of blissful summers and cold winters when soups and sauces warmed the heart, bringing back memories of boyhood days long ago.

Maria Stenger

Now residing in Madison County, Kentucky, Maria Stenger currently has one of the largest collection of Kentucky heirloom tomatoes that I know about. In addition to the three favorites she mentions here, Maria grows and sells seeds of more than fifty Kentucky heirloom varieties.

When I started saving tomato seeds in the late 1990s, I first met Gary Millwood online. Gary's enthusiasm for growing tomatoes was infectious. He shared with me that finding an unknown local heirloom is fun. He shared so many tomato varieties with me that I became spoiled and only planted tomato varieties from Kentucky. Gary and I unearthed over fifty unique tomatoes.

Barlow Jap: This is one of the best for taste. Roy Barlow of Shelbyville, Kentucky, was a teacher by trade, but his great love was heirloom tomatoes. He built a thriving business selling them each summer. After World War II, a friend of his brought him tomato seeds he had smuggled out of Japan. His friend said the seeds were from the best tomato he'd ever tasted. Roy planted the seeds each year after that and developed what started out as "straggly looking weeds" into healthy, disease-resistant potato-leaf plants. The tomatoes were so tasty that they became the main crop in his produce business. His granddaughter, Brita Barlow, has been saving the seeds of this variety to keep her Papaw's legacy alive.

Purple Dog Creek: This very tasty family heirloom with a "purple" color comes from the tiny community of Dog Creek near Munfordville in Hart County, Kentucky. A picnic supper was given for a Martinsburg, West Virginia, minister and members of his church who had arrived to do some home improvements for the low-income elderly people in that area. These tomatoes were served, and they were so good that the man who grew them went home and returned with some seed to share with the minister, who in turn shared with others.

Aunt Lou's Underground Railroad: This is a very productive all-purpose tomato. A black man from Kentucky, traveling through the Underground Railroad, arrived in Ripley, Ohio, with the seeds of this tomato variety. Ripley, where many slaves crossed the river to freedom, is home to Rankin House, a well-known stop on the Underground Railroad and now a museum. The black man shared his seeds with a woman named Lou, who later shared seeds with her great-nephew Francis Parker, who lived in Sardinia, Ohio. Sixty years later, Francis shared seeds with Wilfred Ellis, owner of Ellis's Feed Mill. Wilfred shared them with Susan Barber, who shared them with Kentucky tomato guru Gary Millwood and me.

Lola Wright Choinski

Seed savers find me at farmers' markets, on our website, and at various seed-swapping occasions. Lola's son found me on the Internet. After our telephone conversation, she sent me a long letter about her life in Kentucky. I'll share some excerpts here.

I was born 1-23-1938 at home in Beefhide, Kentucky. The post office was Lionilli. I am now past seventy years old and I have memories of many things from the age of two or three. My father rented our house from his sister, my aunt Vina. She had married Ode Mullins' son Amos Mullins. Old Mr. Mullins had—to us—a very fine house, painted white and two story. Mr. Mullins lived just to the right of the main road up Beefhide at the beginning of another hollow—branch—called Bear Fork. A bear was killed at the beginning of this hollow, thus the name. Old Mrs. Mullins was Nancy, and she was bound to a wheelchair. I do not remember why she was handicapped. Ode Mullins had a huge amount of flat or bottom land on the Beefhide road side and a lot of land up Bear Fork. He donated the land for a one-room schoolhouse on the land facing the creek and road on Beefhide. I attended this school along with five other brothers and sisters. We later moved farther up Beef-hide to live in a log home that was called Miz Ark's Place.

Willard Burke had a large mercantile store close by, and the Beefhide school was just out from his store. My aunt Alka had married a distant cousin, John Wright, who was the main teacher at this three-room school. Only two rooms were ever utilized. We children played in the empty room when the weather was too cold or rainy. The first school I attended was Lionilli. I was in the primer. This school also had three rooms, but only two were put to use. Again, we played in the empty room with inclement weather.

Manuel Burke, brother to Willard Burke, was the head teacher—principal. He also was the postmaster. He had the post office in a corner of his front room/parlor. There were very few families living along the creek that ran from the head of Beefhide to the mouth where this hollow started. I'm not sure if the creek emptied into the Big Sandy or the Kentucky River or another creek that did empty into one of the bigger creeks or rivers. I do know that many, many times a good rain caused the creek to overflow its banks, and there was a lot of damage and loss of life—man and animals. We thought it was exciting that chickens, pigs, dogs, cats, calves, cows, or humans would go floating by. Usually the people had hold of a tree, boards, mattresses, animals as they went by. My dad would tie himself to a big willow tree with the ropes that he used as plow lines and would wade out into the flooding creek to pull to safety people, animals, furniture, or any wood that could be used as stove wood after it dried out. Of course my older brothers helped. Mom often took up the hill to where the barn was to safety. We thought it was a great adventure. Of course we had the animals to take care of.

My dad was a coal miner and a sharecropper also. He was an excellent barber whenever someone needed a haircut. Many, many times he was tossed a nickel or dime after he was through; a shave, with a straight razor, was included. In later life when we moved to Thornton in Letcher County— just six miles from Whitesburg, the county seat of Letcher County—the men would give him 25 or 50 cents. I never knew them to pay him more. He trusted me to give him a cut and sometimes a shave. I did "strop" his razor for him on a leather strap he had just for this purpose.

My father was really a colorful man; he had done so many things in his life. I wish you had been able to know him, he knew so very much about crops, especially when to plant. He used the Old Farmers' Almanac plus signs from nature. There was a time in spring that he called "The Chinook Wind" that was a sure sign to start planting. A lot of the garden was planted on Good Friday. Seed were saved from year to year. Very few were bought. Dad saved corn, popcorn, sweet potatoes, potatoes, sorghum cane, tomato, beans, peas, cucumbers, sweet pepper, broomcorn, pumpkin, cushaw squash, mustard, and turnip. I'm sure I'm leaving out some of the things he saved, but anything that could be saved was saved.

The other things he needed he was able to obtain from one of the stores when he went to town—Pikeville and later Whitesburg. He also saved watermelon and what we called muskmelon or cantaloupe. He did swap some things or sold a lot of seeds to people when they needed them, such as corn, potatoes, or sweet potato—yam—slips. He bedded down sweet potatoes starting the 15th or 16th of January for slips. The beds for the sweet potatoes were made with horse manure on the bottom, next a deep layer of broom straw—sedge grass—that grew wild on the land that was too steep or soil too rocky or poor to use for farming, and a layer of dirt. Sweet potatoes were laid on top, then more dirt, sedge grass, manure, and then a top layer of dirt. The manure caused heat as it rotted, which caused the potatoes to sprout.

I don't know if you know what a "potato hole" is or not. We always had one to keep or winter over any things that could be put in to prolong their time for usage. Most of the time the hole was under the house for protection from the weather and any other danger from varmints or those people who would rob or steal your hard-earned bounty. Honestly, I guess because of World War II the worst came out in people. We had rationing, as I'm sure you know of, and people would steal anything they could get their hands on instead of planting their own crops or working for them.

My father worked for one dollar a day and the horse or mule was a dollar a day. That was from sunup to sundown. We had to get our chores done before breakfast and after supper. The older boys would receive 50 cents each and the girls or younger boys got 25 cents per day. We children never saw any of our hard-earned money. It was paid to our dad at the end of each day. Money was in short supply in that area of the country during the Depression and the war. We were able to trade some rationing stamps for ones we needed. Dad made sorghum—long sweetening—in the fall. Grown just like corn. I remember when I was four years old—I have a picture of the boiling pan, etc.—we made 1,500 gallons of sorghum that year. People came from everywhere for that sorghum, even other states, with their jugs, jars, buckets, or anything to carry the sorghum in. We never owned a car, so we were able to trade our gas stamps for other useful, to us, stamps.

What a time in our history. I don't know anyone who tells stories about this time in our past to others. It, like the growing of crops and the preserving of foodstuffs, has slipped into the "good ole days." I try to tell my children, grandchildren, and great-grandchildren about this time in our lives. I'm lucky that the grandkids and my nine-year-old angel—Chloe Noel—love to hear my stories. She was the first great-grandchild. I've taught her to sew, quilt, garden, and also cook. She is not a child who would rather watch TV but is willing to do anything I show her. Maybe grandparents should take on the task of teaching children about our ancestry and all that it entails because the parents are always too busy trying to keep up with the Joneses or getting all the material things, that they don't teach the children what the most important things in life are. I wish we could have realized that all that is important is morals and the act of helping. I guess I'm just wishful thinking.

I do think that if children had more good and interesting things, such as planting a garden—watching it grow and picking the bounty—they would not be getting into so much trouble. I've taught my ten grandchildren about planting and growing since they were in a stroller and they all were interested, especially the eating of their own vegetables. I'm sure that as teenagers and young adults they have laid this knowledge aside, but it will always be with them. Nothing is as great as to hear one of them ask me to tell them another story or to repeat one I have told them over and over yet again.

Our lives were full; of course most of it was work. We

may have gotten hungry but we never went hungry. We usually worked up 30 or more bushels of peaches and just about as many beans. Late at night until 11 or 12 p.m.—out on the porch—both boys and girls were peeling peaches or stringing and snapping beans. Mom usually did up 800 quarts and half gallons of canned food, for our mealtimes or to preserve. We always grew many beans, as they could be canned, dried, or pickled. We called the dried beans "shucky beans" because they sounded like dry shucks when shaken. Mother always stored them in bleached white feed sacks. Apples, peaches, and pears were also dried and placed in these sacks, hung on nails behind the wood-burning cookstove where they would stay dry and the bugs or mice could not get into them.

Looking back now, I don't know how my parents were able to do just part of the things that they did. We would sometimes have eight to ten bushels of beans to be strung, broken up, and dried or canned at one time. Pickling beans was part of our process for preserving some of them. After we had them "worked up," Mother would cook large kettles of them in plain well water, cool them in water from the well, and place them in 20 gallon or larger crocks with a mixture of ¾ cup of kosher salt added to a gallon of cold well water poured over them. It took quite a few gallons of brine to cover them by 4 to 5 inches, sealed tightly by at least 4 layers of white feed sacks tied tightly around the top, and placed in the can house, which was partly dug into the side of the hill. Or they might be put into one of the bedrooms for four to six weeks, depending on how sour you wanted the beans to be.

When they were finished pickling she would remove the covers, skim the mold or mother from the brine, and then put them into capped canning jars and then in the can house for the winter or until they were eaten up. Mother also did pickled corn on the cob and mixed pickles the same way with the same brine—fresh, of course—as the beans.

She always grew little white half-runners, goose craw, little white greasy beans, cornfield beans, and fall beans. They were all pole beans. The fall beans and cornfield beans were planted in the corn, but the goose craw beans were stuck with limbs or brush from the woods. White half-runners were never put up on poles; they grew in a tangle and supported themselves. Mother always had a hoe near her because of the possibility of poison snakes. We also had cows, pigs, chickens, ducks, geese, guinea hens, mules, and horses to take care of. Dad grew a white corn that people around us called the George Wright corn, for this was the only corn he would grow. He was very picky about his corn seed that he saved from year to year. We used this corn for grinding for our own use plus feed for all of the animals.

I recall many, many times we shelled the corn that Dad had selected from the corn crib, shucked, and brought into the house. All of us were pressed into service to shell four to six large, heavy-material sacks that held two bushels each of shelled corn. Dad would take the sacks of corn to a mill to be ground so we could make cornbread or cook mush for our meals. A fourth of what he took to the mill was used to pay for the milling. We never went hungry. I know very few people today, if any, who could live like this. Sometimes I wish we could still be preserving our old ways instead of reading about it or watching films about the ways of life back then. We certainly had a lot less crime, since children were kept busy helping the family survive.

Dad planted his crops by the signs of astrology or what

was written in the Farmers' Almanac. I don't know how or when he started doing this, for he was doing it before I was born. As close as I can guess, he bought the almanac in town—Pikeville—when he went to get our other supplies such as flour, salt, and sugar. He rode a horse, led a mule when he went into town. I do know that we always started planting the garden on Good Friday. Anything that grew in the ground, such as potatoes, were planted when the signs were in the feet. Nothing was planted when the signs were in the bowels. When the leaves on an oak tree were the size of a mouse ear was the time to get the corn planted. If we were not hoeing the corn for the first time by May 1, we would not have a good return on what was planted, according to Dad.

Honestly I never recall us not having a successful season. We were always asked by people that didn't want to work for foodstuff, even the scraps from the table. Dad would tell men he would give them the use of his mule and the seed to plant, but for some reason they always had an excuse such as lumbago, rhumatiz, or other strange afflictions. During World War II, Dad and my brothers had to patrol our fields with guns, as people would steal the food right out of them. Also, we lost many of our chickens, and even the cows would be milked during the night. That is when we got the guinea hens, for they make a terrible racket when disturbed.

The corn Dad grew was already in use when I was born. I don't recall it being called anything but George Wright corn. It was white and a very large size to the ear, at least 12 or more rows on an ear. It sure was delicious when picked young and boiled for roasting ears. People came from all over to buy seed from him each year. I don't know why they didn't save some themselves. Guess they didn't want to do the work. Mom always made hominy in the early spring from this corn and canned it also.

I left home two days after graduating from Whitesburg High School and went to Detroit, Michigan, where ten members of my family, brothers, sisters, and in-laws, were living. A sister and her husband came to my graduation, as I was the first to complete a high school education. I had hoped to go to college at Berea or Caney Creek, where you could work your way through school, but Dad would not give me permission to do so.

May 19, 1955, was my graduation. I was in Detroit on May 21 and started a job at Standard Accident Insurance Co. on May 23. I lived with my sister for a year; they moved to Southfield, Michigan, but I roomed at the Pricilla Inn, just across a park from my job, so I would not have to pay bus fare. I sent $20 a week home to help Mom and Dad and the younger children of my family. I married two and a half years later in 1957 and am still married to the same man. We moved to North Carolina in 1993 because of my husband's job.

I have always had a garden since we were married. We bought a home in Troy, Michigan, before we were married. My husband's family were Polish farmers in Warren, Michigan, and the near area. Our brother-in-law had a tractor, so we were able to get the ground plowed. I have always kept seeds from anything I could, just to see what would grow. I will admit it has helped to have a garden and more as to the food I preserved. I still do a lot of what Mom did, but I have a freezer and I cook on an electric stove. I have tried to impart what I know to my five children. My son John has a small garden in his backyard. My little great-granddaughter is so very interested in what I do; she has helped me plant and pick the garden since she was two years old. She is an A+ student at school and can really tell about her "farming" to her classmates. She calls it a 'venture since I started teaching her when she was only three days old. Of course that is another story and everyone thinks I am nuts, but when they hear her talking, they are stunned she knows so much.

I was waiting before I sent this, waiting for my sister Olivia—who lives in Oklahoma—to send me some of the beans she and her husband Jay call "Goose" beans. Jay's family traveled overland in covered wagons and settled in Kansas—lived in soddies and farmed. I don't know—neither does he—just when he started these beans. He is in his mid-seventies. I was hoping they would be different, but they seem to be the same ones that my brother's neighbor, who lives in Dana, Kentucky, gave him to send to me last year, and also just like the ones from you. I'm sending some for you to compare. I'm also sending you some of the cornfield beans that have been in my family as long as I can remember. I think they are different than the ones I ordered from you. They take longer to grow but they are really, really good when cooked. Also I'm sending some of the white half-runners from my family. I put brush in them and they will grow very tall. I'm usually picking half-runners by the middle of June or earlier and am still picking when we get a killing frost in late October or later here in North Carolina. One year I picked on December 4 and then they got killed by a heavy frost that night.

Ruth and Rudy Thomas

I first heard from Ruth Thomas of Albany, Kentucky, on July 14, 1988, in response to the previously mentioned article by Judy Sizemore in the *Rural Kentuckian.* I knew her son Rudy well. A Berea College graduate, he had worked for me for several summers in the Upward Bound program in the early 1970s. He became a teacher in Clinton County, one of the six counties served by that program.

> Dear Mr. Best and family,
>
> I know you don't know me. But you know my son Rudy Thomas. I read about you and your family in the *Kentucky Magazine.* How you garden and sell vegetables. We raised stuff like that and sold to send nine kids through school and milked cows and tobacco and hauled hay till midnight many a time. But we couldn't of done it without our family's help.
>
> Now what I'm writing about is this. I've planted a lot of beans but I never heard of a greasy bean seed. What in the world is that kind of bean? Is it like the partridge head bean? I thought the beans they show in this book kind of looked like them. Well I'll close, sorry about so many questions. But Rudy said he didn't think you'd mind. Take care and good luck with your gardening.

Mrs. Thomas's observation was accurate: the greasy beans in the photograph did look somewhat like the Partridge Head bean that is so prevalent in Clinton and nearby counties in both Kentucky and Tennessee. The Partridge Head bean is a fast-growing, full bean with an excellent flavor, much like greasy beans—the difference being that it doesn't have the shine of greasy beans. And her observations about the importance of family help with gardening and farm chores are very important.

Rudy Thomas taught for many years and then became superintendent of the Berea Community Schools. He is currently the director of the Upward Bound program at Lindsey Wilson

College in Columbia, Kentucky. He also operates a small press, the Old Seventy Creek Press, in Albany. He has authored many books of fiction and poetry and is well known for his writings on Appalachian culture.

Rudy, who gardens and saves seeds, remembers well the struggles of his parents and siblings on their small farm. Some years after her death, I shared his mother's letter with him for the first time, and he shared the following remembrances of her and the rest of his family. He was very touched by his mother's letter written twenty-two years earlier.

I remember when and how my mother came to be a scavenger of heirloom seeds. It was in the spring of 1953 that I went with my father to the country store and garage where he worked. We went there to pay our bill for foodstuff that had been bought on credit, to be paid for when our tobacco crop sold. It turned out that our bill exceeded the proceeds of our cash crop. It was at that moment that our lives changed. There were four of us kids at the time and our two parents. I say our lives changed because my father told my mother that we could not live year after year and go in the hole like we had done the previous year.

There was no money for seeds or fertilizer, so my mother became the seed gatherer. In the traditional way that farms and gardens had been maintained in Ohio, my mother's home, and in Kentucky, my father's home, we maintained ours.

Mother went in search of what I did not know to be heirloom seeds, but I understood that our community had gone through the Great Depression without any money in a bank but with smokehouses full of cured meats and cellars full of canned fruits and vegetables.

The seeds she began to gather up had been chosen, kept for planting, and handed down from one family member to another for many generations. It was out of necessity, rather than an enthusiasm for cultural or environmental matters, that she began her quest.

To keep from charging the cost of fertilizer at the country store, we hauled manure from the barn. Horse manure was easier to handle than cow manure, I soon learned, but the easiest manure we spread that year was chicken litter from a neighbor's brooder houses.

From my father's family, my grandparents contributed Kentucky Wonder pole beans that my grandfather's parents had brought with them from Tennessee when they fled that state during the Civil War. My mother also got half-runner beans that my grandmother had strung up on strings to dry for use in soups. The half-runner beans were also contributed by my aunt Golda, along with another bean that may have been the Partridge Head bean that my mother loved, or perhaps greasy beans.

I got an education early about pole beans. My mother's plan that first year was to can enough green beans to last a year. Since the half-runner beans would mature about a week earlier than the Kentucky Wonder pole beans, she decided we would eat the half-runners and she would can the longer pods of the pole beans. Those Kentucky Wonders produced 123 quarts.

On Salt Lick Creek, my father and I went to his cousin's house and bartered bonded whiskey for river canes and yellow Eight-Row corn, enough to plant several long rows the length of our garden—a plot as large as the tobacco patch of half an acre. We also got a handful of white Hickory Cane corn seed. My father told me how he used to take a

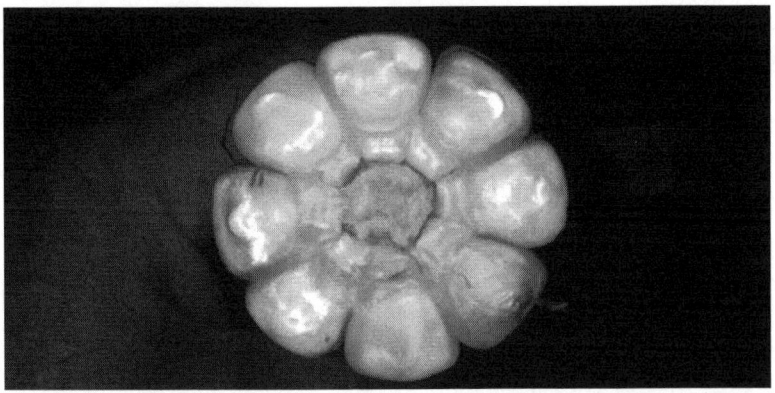

A cross section of white 8-Row corn. I think Hickory King and Hickory Cane are the only two varieties of heirloom 8-Row corn still being grown. Kentucky's Native peoples also grew 8-Row corn. (Michael Best)

turn of Hickory Cane down the hollow to his uncle's mill, and that is how they got their meal for cornbread. He also told me that his cousin's husband used the white Hickory Cane kernels to make his shine.

I remember my father telling me about the wheat seed my great-grandparents planted, grew, harvested with a wheat cradle, and used to make flour. A Cherokee Indian who lived in the hills and was an escapee from the Trail of Tears had given them the seed. He grew the wheat in the river bottoms. The Cherokee and my great-grandfather hid out in those hills together during the Civil War—my grandfather to keep Champ Ferguson, the notorious Civil War guerrilla leader, from finding him, and the Indian because he did not know where home was. A neighbor's son was killed near Creelsboro during the war.

I know my mother got tomato seeds on cheesecloth from other family members, and we sowed them along the poles of the tobacco plant beds. There were red, pink, and yellow tomatoes in that garden, and the taste of them made me a lover of tomato sandwiches, first on homemade biscuits and later on "loaf bread," as we called store-bought bread.

We planted Irish Cobbler potatoes and a red potato that my grandfather loved. I don't know if it was Pontiac Red or not. Mother did not care for the red potatoes, saying they tasted earthy. She did not keep red seed potatoes for future gardens.

From the same family sources, Mother brought home sweet potatoes, red and white, to sprout in water in the kitchen windows. Those slips that took root in the windows grew in a ridged row almost the length of the garden. Family-saved cucumber seeds were used to complete that row and another long row.

Our gardening venture was such a success that after that first year, we planted larger and larger truck patches. My mother's interest in seeds continued. I know she came up with greasy beans, rattlesnake beans, Fordhook limas, a purple speckled butter bean, and a world of other seeds that I have long forgotten about.

I liken us to the Archaic Indians who no longer hunted bison, mammoth, and mastodon as the Paleo-Indians did. The Paleo-Indians had developed projectile points such as the Clovis, the Beaver Lake, and the Cumberland, which helped bring extinction to those large beasts. For the next 7,000 to 8,000 years, the Archaic Indians depended on simpler tools as they became food gatherers and survivors.

We became a seed-gathering family as well as survivors, and Mother became the keeper of seeds—all of them open-pollinated varieties, including Granny Brown and Effie beans and Cow's Horn okra.

Harriette Simpson Arnow

In her well-researched and highly regarded book *Seedtime on the Cumberland,* Harriette Simpson Arnow talks about how beans were planted, harvested, eaten, and preserved. Yet the word *tomato* does not even appear in the index of that book. Beans have long been a significant part of the Kentucky and southern Appalachian diet. The "love apple," or tomato, wasn't eaten until much later, after it was determined not to be poisonous. Here's an excerpt from Arnow's book:

> Those (vegetables) most commonly grown on the Cumberland were the old standbys on any frontier—beans, peas, turnips, and cabbage. These like the pumpkins which sometimes reached a weight of 140 pounds, grew well on rich new ground soil, and had the advantage of keeping in such storage as a first settler could provide, though second crops of turnips and cabbage were often started in late summer, and not harvested until late in the fall. Many, particularly the Germans, made their cabbage into kraut, a most useful food to any family before the invention of canning, for it could for long periods be kept in kegs or barrels. They may also have made turnip kraut as did a neighbor of my childhood, quite a good dish, as was also green beans put through a fermentation process, but known as pickled beans instead of bean kraut. More enjoyable than either to the pioneer with neither apples nor pears to store for winter, were the raw turnips and cabbage.
>
> Beans and peas were favorites and could be grown among the corn, while those not eaten green were stored; two bushels of peas bringing only $2.50 in 1803 as compared to 83 cents for only one bushel of sweet potatoes; these, sometimes known as "Carolina potatoes," had long

been a favorite in the south for roasting in the ashes, frying, or mixing with milk, eggs, and spices to make pie filling. Some families would have used a variety of pea, known as the salat pea, still a few years ago grown in the back hills from handed-down seed. These, instead of being hulled, were prepared and cooked with meat much as were green beans.

These last, boiled with bacon, are still a favorite on the Cumberland, sometimes cooked in an iron pot in much the same manner as Mrs. Jocelyn, south of Nashville, cooked those she served her visitor, Francis Baily, in the summer of 1797. Mr. Baily didn't relish the dish of bacon and beans too well, but I can think of no better eating, if prepared by the old ways handed down. The preparation of a mess of beans was just about as important a job as the making of good butter. Much depended on proper beans; the favorite varieties were cornfield bean—the speckled goose-craws or the brown Octobers—and these must be well plumped out, with a few brown-hulled ones to be shelled, their soft, bright seed to be cooked with the green ones.

Beans for the next day were picked late in the afternoon into a basket or a tucked-up apron, and when the early supper was finished, the woman, often a grandmother, if she could see well, sat with them in her apron, broke off the ends, pulled off the strings, and broke them into pieces. Next morning they were put into a kettle, iron of course, and cooked during breakfast getting; after breakfast they were taken to the spring house and changed and washed through two or three waters. They were put on the fire again, outside in hot weather, and this time a large chunk of cured hog meat, jowl bacon would do in beans, sliced down to the skin, but the lye-water-scrubbed skin left was put in. Beans and bacon cooked along till noon; you could eat a few for dinner, but they were better left simmering along till supper.

Sometime along before supper some Irish potatoes would be peeled, quartered, and buried in the beans; a little later some roasting ears, and last the okra pods on top. All this with cornbread, sliced cucumbers, pickled beets, red pepper relish, with some fried meat and gravy, and the usual wild honey, preserves, pickles, slaw, and sweet potatoes, baked in the ashes, made quite a good supper.

Dried green beans, known as shuck or fodder beans, needed an even longer cooking but in winter made a fair substitute for fresh green beans, and in drying them in summer the old ones followed much the same recipe as that given in a Nashville newspaper of 1811, save they did not put in a teaspoon of sugar. However, the main bean in winter was the dried bean, almost always boiled, for beans boiled with bacon and less often ham hocks, were to the Cumberland partly what the baked bean was in New England, though in that country of abundant meat, beans and peas together were never as important as in New England.

Most Cumberland housewives hulled their beans as did my great-grandmother. She first boiled them in weak lye water, then carried them to the spring branch where, like hominy, they were washed in a basket, until the cold water on the hot swollen beans made the hulls slip with little trouble. Her husband would not have eaten unhulled beans, and certainly she would never have served them. The English cookbook of 1701 directing housewives to hull their beans had not been published when her people came to America, but the old ways lived from mother handed down to daughter. (411–13)

Wayne County traditional gardener and seed saver Frank Blevins with his homemade corn sheller.

In her earlier, widely acclaimed novel *The Dollmaker,* Arnow was always cognizant of the role played by foods in the lives of her characters, particularly Gertie Nevels, the book's main character. Gertie's husband, Clovis, was a mechanic, a "tinkerer," while Gertie was a farmer and gardener with highly honed skills.

When the Nevels family went to Detroit during World War II, Clovis found a world that was much to his liking, working in the industries that made machines for the war. Gertie, however, was out of her element, forced to be dependent on Clovis's wages to pay for store-bought food that was far from fresh. Gertie was always thinking about what she would be doing at home—going to the root cellar for apples or potatoes, frying fresh meat from a hog she had butchered herself, or getting ground ready to plant potatoes, corn, and then beans, always adhering to the signs taught by her father. As her time in Detroit continued, Gertie was painfully aware of what she was missing at home: gathering roasting ears from the corn, picking plump beans from the vines climbing up the cornstalks, or frying eggs fresh from the nests of her chickens.

Harriette Arnow was so tuned into the seasons, based on personal knowledge, that she could write eloquently about what Gertie would have been doing back home at any time of the year.

8

Practical Tips for Growing and Saving

We gardeners and farmers love to sit around and tell tales about our successes and sometimes our failures. We are always talking about something we tried this year that worked well, or maybe it didn't. This is what makes gardening and farming so fascinating and challenging. Every piece of land is different, with different soil and a different orientation to the sun. So what works just great for me may not work so well for someone else. Then, of course, there is always old-man weather. No two years are ever the same. And once gardening has skipped a generation, it is unfortunately necessary to start from scratch: knowledge passed on for hundreds of years has to be relearned, accompanied by trial and error.

The best thing to do is make good notes every year: what you planted, when you planted it, how it grew, what the harvest was, and of course, how it tasted! Here, I offer a few practical tips from my perspective.

Cornfield Beans

With beans, it is good to know that a few things have happened in the last few decades that have forced traditional practices to

Turner Cornfield beans, grown on a bean tower, are taller than the twelve-foot stalks of Hickory King corn.

change. Traditionally, cornfield beans have been planted with corn so the cornstalks could provide the "poles" for the bean to climb. But with the advent of modern hybrid varieties, the cornstalks are too weak to support the bean vines. At best, hybrid cornstalks, both sweet and field, can support only one or two ears of corn and will collapse under the weight of bean vines. Therefore, most people who are serious about growing climbing heirloom beans use poles or a trellis to support the bean vines, or they grow them on heirloom varieties of corn such as Hickory Cane. Trellises should be only as high as you can reach without using a ladder to pick the beans.

Another way to support bean vines is to construct a bean

Frank Barnett's bean tower. It's important not to plant heirloom beans too thickly. Frank gets outstanding growth by not crowding the vines, planting only a few seeds around the tower. (Frank Barnett)

tower made from a stout pole with a bicycle tire rim on top. Strings are attached around the perimeter of the wheel and then attached to the bean vines on the ground. These bean towers need to be at least ten feet tall. You can use a stepladder to reach the beans growing higher than your head. Bean towers are an excellent way to save seeds and "get a start" if you have only half a dozen or so seeds. The tower, allowing for ample vine growth, makes it possible for the maximum number of pods to form and to produce the most seeds from the smallest number of plants.

Bill Best's bean vines just beginning to run and bloom.

Cornfield beans need to be planted at a rate of two seeds per eighteen inches. The two seeds help each other break through the soil at germination and then "spread their wings" as side branches quickly develop on the main stem. Virtually all modern seed companies give bad advice when they promote the sowing of bean seeds at a rate of every two inches or so. Mechanical planting devices also space them far too close together. Of course, commercial seed companies are in the business of selling seeds, not promoting good vine growth or growing quality beans.

Once the beans mature on the vines, it is important to save the seeds at the appropriate time; otherwise, the seeds can be damaged by weather conditions. If the weather is clear and dry at the time of maturity, the bean pods can be left on the vines for several days until the pods become dry, at which point they must be removed from the vines. If it is rainy when the bean pods mature, it is best to remove the pods and spread them out in a dry area to complete the drying process. A greenhouse or high tunnel works well, assuming the pods are spread out on plastic on the ground or placed on greenhouse benches covered with bedsheets or some other cloth.

Drying can also be accomplished by spreading the pods over the floor of any dry room in the house or barn. The most important thing is to prevent the pods from getting wet during the drying process, as the seeds will sprout or mold within the hull. If this happens, the seeds will never sprout in the ground and won't be good to eat either, even as dry beans.

After the seeds have become dry, shelled out, and hard to the touch, it is important to remove the disfigured or insect-damaged seeds from the batch. If some seeds are a different color than the others, these seeds can be planted separately the following year to see if they breed true. If they do, then you might have discovered your own bean variety. This is the process by which we have developed thousands of varieties of heirloom beans.

Heirloom Tomatoes

To achieve good production with a minimum of rot and sunscalding, tomatoes need to be staked or trellised. It is also possible to use cages made from concrete reinforcing wire, fencing wire, or any other wire that can withstand considerable weight. This gets the tomatoes off the ground and provides plenty of shade to prevent sunscald of the ripening fruit.

Tomato seeds can be saved in several ways. One of the traditional methods is to let the tomato ripen completely, even to the point of beginning to rot, and then remove the seeds with a spoon and spread them on a piece of cloth or paper to dry. Some people spread them out on a paper towel, let them dry, and then plant the paper towel and seeds together in potting or germinating soil.

A far better way to save tomato seeds is to use the fermentation process. The tomatoes are allowed to overripen to the point of beginning to rot and then quartered or cut up so that the seed cavities can be scooped out and put in a bucket or some other container. You can do this with one tomato or with many, depending on the number of seeds you want to save. The tomatoes are then stirred one or more times per day for three or more days until the mixture is soupy. Fungal growth will appear on top of the mixture as fermentation takes place, but that is no problem. During stirring, the seeds dislodge from the gel and sink to the bottom of the container. Water is then poured into the mixture, allowing the pulp and the bad seeds to rise to the top and flow over the side of the container. The good seeds sink to the bottom. Once the water becomes clear, pour what's left in the bucket into a finely meshed strainer. Only the seeds will remain in the strainer. Then spread the seeds out on a flat surface, such as a slick paper plate, to let them dry. My own preference is to spread the seeds on wax paper and put it under a slow-moving fan until the seeds are dry, which usually takes no more than twenty-four hours. Once the seeds are dry, you can scrape them off the paper with your finger and separate any that might be stuck together. I then put the seeds in a tightly sealed plastic bag, dated and labeled, and store the bag

Brian Best stringing high-tunnel Vinson Watts tomato plants as high as he can reach.

at room temperature, making sure it is not in direct sunlight or in a hot part of the room. Using this method, I have had good luck germinating tomato seeds saved for up to ten years.

When sowing tomato seeds, it is important not to plant them too deep—half an inch is adequate. Keep the soil mixture warm and moist but not wet. Most tomato seeds germinate within four to seven days. They need a lot of sunlight at this early stage to prevent the plants from becoming elongated and weak. Commercial full-spectrum grow lights placed close to the germinating plants work best for producing early transplants. The plants should be ready to transplant within six to eight weeks. As soon as suckers appear on the plants, break them off below the first bloom clusters, which will now mature much earlier. Suckering also keeps most of the foliage off the ground, helping to prevent disease.

9

Heirloom Favorites

I have been collecting heirloom vegetables for more than fifty years. I am often asked for my favorites, and that is a hard question to answer. I have to admit that I have more than one.

Favorite Beans

When two of my granddaughters were toddlers, they instructed their mothers to serve them "real" beans. They meant they wanted to be able to see the bean seeds on the spoon before putting it into their mouths. Without knowing it, they were very much into tradition. They wanted what mountaineers call "full beans."

Likewise, I want my beans to be full—tender and full. During all my years of collecting, growing, eating, and selling heirloom beans, I have refused to collect any green bean that must be eaten before the seed forms—what one might call a protein-free bean. Virtually all commercial beans that are meant to be eaten as green beans are too tough to eat once the seeds appear. They have to be tough to withstand the rigors of mechanical harvesting. Further, most tender beans are climbing beans. It is very unusual to find a bush bean that is both full and tender, but there are some.

All of this having been said, here are some of my favorite beans in their respective categories.

Beans ready to eat! Full of protein and ready to stick to your ribs.

Cornfield Bean: My favorite cornfield bean is the Barnes Mountain Cornfield bean from Estill County, Kentucky. I first obtained it when Ott McMaine of Waco traded me some seeds at the Lexington Farmers' Market in the 1970s. It is a large white-seeded bean that has a mild flavor and remains tender throughout its growth period.

Greasy Bean: It's hard to find anything I don't like about any greasy bean, but I have to choose the Pink Tip Greasy bean as my favorite. This bean from the Bethel area of Haywood County, North Carolina, is also a favorite among my customers. As this bean matures, it develops a pink tip on the end of the pod. I know of no other pink-tip greasy bean. As is the case with other greasy beans, the seeds separate from the hull when the bean is cooked.

Greasy Cut-Short Bean: All cut-short beans have seeds that are crowded in the hull, giving them a high percentage of protein. My favorite carries my mother's name: the Margaret Best Greasy

Cut-Short. My mother shared the seeds with me many years ago. It is also a favorite of my customers.

Fall or October Bean: Although fall beans sometimes have tender hulls, most people eat them as shelly or dry beans. They can be cooked and eaten at the time of shelling or used later as dry beans. Most traditional gardeners of the southern Appalachians grow one or more varieties. My wife, Irmgard, prefers fall beans over green beans. We have seed for only half a dozen or so varieties, and she likes them all as either shelly beans or dry beans. My favorite is the Roger Newsome Fall bean from Letcher and Perry Counties.

Half-Runner Bean: I now have at least twenty varieties of half-runners, and my favorite is the one we call the Non-Tough (NT) Half-Runner, so named because it remains tender throughout its growth period. It is good for eating fresh, for canning, and for making into leather britches. This bean originated in Breathitt County, Kentucky.

Favorite Tomatoes

Tomatoes tend to be color-coded, and most people grow or buy their heirloom favorites based on color.

Pink: Pink tomatoes tend to be high in acids and sugars— what many refer to as "old-fashioned" flavor. The pinks are my personal favorites, and the Vinson Watts tomato is my favorite one of all.

Red: Red tomatoes tend to be high in acids, and most commercial hybrids are red. Even the earlier hybrid red tomatoes, such as the Ramapo, have good flavor. Heirloom red tomatoes are also high in acid and are pleasing to a lot of people.

My favorite red tomato is the Zeke Dishman, a very large and tasty tomato that often weighs over two pounds. It was developed by Zeke Dishman of Windy in Wayne County over several decades.

The Willard Wynn Yellow German tomato. (Dobree Adams)

Another red tomato with exceptional flavor developed over many years is the Mary Rose McMurray tomato from Harlan County. It is a very large Roma type, but unlike the commercial Roma tomatoes that have little if any flavor, it has an excellent acid flavor and is a good eating tomato in addition to being good for canning and sauce.

Yellow: Yellow tomatoes tend to be high in sugars. Although they are thought to have a lower acid content, the sugars simply overpower the acids and give the tomatoes a sweet flavor.

Irmgard's personal favorites are any of the German Yellow varieties, which have red stripes or blushes, because of their large size, texture, and sweetness. My favorite among the German Yellows is the Willard Wynn, a Kentucky heirloom developed over several decades, first in Harlan County and later in Rockcastle County.

Claude Brown's Yellow Giant, my favorite yellow tomato, is

Willard Wynn, a longtime seed saver originally from Harlan County. For many years we have grown and sold his namesake, the Willard Wynn Yellow German tomato.

actually a deep orange color and can weigh three pounds or more. It was developed by Claude Brown of Pike County, Kentucky.

Other Colors: Tomatoes come in many other colors, including green-when-ripe, brown, black, and purple. Their various sugar-acid combinations give them distinctive flavors.

Of these tomatoes, I prefer the black ones. My favorite black tomato (and a favorite of my customers) is the Blackberry, a large and tasty tomato weighing about twelve to sixteen ounces. The original seeds were given to me by John Allen of Cartersville, Kentucky. He wasn't sure of its origin.

This Blackberry tomato, one of the best black tomatoes, is from the late John Allen of Cartersville in Garrard County.

My favorite green-when-ripe tomato is Aunt Cecil's Green, a Kentucky heirloom with good flavor. This variety develops a yellow tint on the blossom end as it becomes fully ripe. It has a very good flavor but doesn't have a long shelf life. It sells well at farmers' markets.

Tommy Toe: Modern tomato breeding has given us a lot of small, very sweet tomatoes that go by the name "cherry" or "grape" tomatoes. What is missing is tomato flavor. We now grow primarily the old-fashioned tommy toes, which have a very rich tomato flavor. One pink tommy toe, measuring about an inch in diameter, appeared some years ago as a mutant of the Vinson Watts tomato. We named it the Robe Mountain Tommy Toe, after the mountain behind our house. Another black tommy toe is a true breeding mutant of the Robe Mountain Tommy Toe. We named it the Basin Mountain Tommy Toe, after the next mountain over from Robe Mountain. Both have excellent, though different, flavors and sell well at the Berea and Lexington Farmers' Markets. We also sell seeds of both tommy toes on our website.

Trays of Willard Wynn Yellow German and Vinson Watts heirloom tomatoes ready for the farmers' market. (Dobree Adams)

10

Growing and Sharing Today

Julie Narvell Maruskin

Although Julie Maruskin tells people she's a librarian, not a gardener, she's descended from a long line of gardeners. Her parents came from Harlan County, moved to Paris, Kentucky, and then to Bell County, and finally retired back to Paris. They gardened all their lives, and Julie developed a love of plants and gardening as a child. Now director of the Clark County Public Library in Winchester, she has developed an outstanding and popular program that teaches gardening and seed saving in a number of counties through the library system. At Julie's invitation, I have given talks about seeds at her program for a number of years (that's where I first met Frank Barnett).

Julie's husband, John Maruskin, is also a librarian and a lifelong gardener from western Pennsylvania. He has developed handouts about planting in accordance with the "signs"—the phases of the moon. For example, plants that are grown aboveground should be planted when the moon is waxing (becoming fuller), and belowground plants should be planted when the moon is waning.

Julie's family history is a good place to start her story.

Julie Narvell Maruskin, director of the Clark County Library, teaches gardening and seed saving throughout the library system.

To begin with, this was all my mother's fault, and I reckon it's also the fault of a man named Grover Hughes.

My mother, Sonia Mae Wilson Narvell, was born in Harlan County, Kentucky, but her mother's family were the Whitley County Pridemores and Davises, farm people who grew a tremendous variety of fruits and vegetables, including at least three varieties of peaches in a pristine orchard set up on the highest ridge on their property. Although the Pridemore farm prospered on livestock and food crops, including corn and oats, it also boasted a very fine secret patch of burley tobacco grown by my mother's great-grandfather, John "Free Sugar" Pridemore, whose Baptist son William did not approve of smoking. "But," as Free Sugar told my mother, "what Willie doesn't know won't hurt him a bit."

My mother never did tell on her great-grandfather, but she remained mightily impressed by everything grown on that farm and in the gardens her parents, Arles and Nola Wilson, grew around their own houses in Harlan to augment Arles's slim coal miner's pocketbook. Apparently, there was one summer where the dining room window in their rented coal-camp house was completely shaded by tall stands of sweet corn, simply because my grandfather was not afraid to till up a grass yard when his seven children needed food on the table.

Being born late and last, I saw my grandparents only a handful of times, but my mother made sure I knew who they were, and she did it largely through vegetable gardens. I saw her favorite brother Floyd Wilson less than five times that I remember, but I never see someone slice a tomato without hearing my mother tell me that Floyd would always ask their mother for "just the flower end, Mom, please." I never ever see a stand of corn without remembering the story about my mother's sister Helen wearing my mother's new Sunday dress and my grandmother's best gloves as she stood blissfully hoeing the garden. And I'm sure to plant my pole beans by the correct moon sign, since my mother loves to tell the story about the time she decided to plant her beans when she darned well pleased—and her mother said, "Now, Mae, I don't think that'll work out well, but you go ahead and we'll see." Sure enough, my grandmother's beans bore their usual heavy burdens of snaps, and my mother's beans flowered to beat the band and then promptly dropped every blossom without making a single bean.

For me, my family history was all scribbled down in phantom garden rows stretching back well into the nineteenth century, and the fruits and vegetables my mother grew or purchased all elicited stories that were fed to us at every meal.

When I was ten, I was the last child at home, and my mother and father moved from Paris, Kentucky, to Middlesboro, down in the elbow crook of southeastern Kentucky. It was a different world for me, but it was very close to home for both my parents, and it wasn't long before my mother was sizing up the local farmers' markets. And it wasn't long before she ran into Grover Hughes, because Mr. Hughes was halfway between our house on Cumberland Avenue and town.

Grover, you see, was possessed by beans. Although he was a longtime Middlesboro resident, he had become a world traveler by the 1970s, and he searched out beans wherever he went. And he grew enough beans to sell them for next to nothing in front of his house on Cumberland Avenue for about eight weeks out of the year.

I tell about Grover's beans now, and people think I'm telling big garden lies. Gardeners as a group are far more prone than fishermen to not let the truth get in the way of a good story. "Lord, that tomato was this big! Here, ask Junior! We canned a quart offa every tomato, I swear to you. THREE quarts offa the biggest one—it was as big as little June Marie's head, and every plant loaded so's you couldn't see any leaves. Wasn't it, Junior?"

However, it is quite true that in Middlesboro, Kentucky, in the late 1970s, my mother was buying a different kind of exotic heirloom bean every few days. She would go down to Grover's, where they would start trading vegetable lies, and she would come home exhilarated, bearing a bag of mysterious beans. Some of them were nearly silver, the color of a luna moth's wings, they were so light a green. Some of them were dark with dark maroon and blue-purple stripes. Some were yellow and curled as tight as a pug's tail; others were blue-green and eight inches long; and some were the length of my little finger, with seeds as small and white as grains of rice. Each of them had its own beautiful and distinct beany aroma and flavor.

So, you see, by the time I moved to Massachusetts for a few months in the 1980s and then later to southern Indiana, my mother and Grover Hughes had pretty well ruint me. I couldn't find farmers' markets like the ones I knew in Kentucky and Tennessee. And when I did find a market, the growers didn't want to talk about the vegetables. They just wanted me to buy the limp, rubbery beans and the equally rubbery tomatoes and get on home. And more shocking than their attitudes were the vegetables themselves. I had never had a bean, not once in my twenty-plus years, that I could tie in a knot before cooking it.

I figured out that if I wanted anything that tasted like the food I knew, the food I had been raised on, I was going to have to grow it myself. Beans were no problem, but tomatoes? I had never started them indoors before. I went looking for seedling varieties and found a handful for the first garden I grew as an adult. The tomatoes were hybrids, and I don't remember their names; I remember only that they were nearly as lifeless as the ones in the supermarket.

So the next year I started looking for catalogs—and this was pre-Internet, so it wasn't all that easy. Also, I was not a millionaire, and this was back when you had to pay—sometimes upward of $5 each (my hourly wage at the time)—for catalogs. Plus you had to order them by mail, so you were lucky if you got them in six weeks. After perusing the advertising sections of *Mother Earth News, National Gardening,* and *Organic Gardening* at the Monroe County Public Library in Indiana, I came up with some candidates, and by one of the best pieces of luck I've ever had, I managed to procure a Southern Exposure Seed Exchange catalog. I started reading it the day it came in the mail and didn't close it again until three o'clock the next morning.

If you don't know Southern Exposure, I can heartily recommend the company. It was started by Dr. Jeff McCormack, a biologist and, like me, a person deeply in love with the sheer diversity of garden vegetables. (The idiot who coined the phrase "garden variety" as an expression of something run-of-the-mill should be dug up and shot, in my opinion. Maybe twice. There is no such thing.)

In the Southern Exposure catalog I found stories about the plant stock they offered and the people who grew them. And the plants! Oh, there were orange oxheart tomatoes (Verna Orange) and pink beefsteak tomatoes (Radiator Charlie's Mortgage Lifter) and yellow paste tomatoes (Yellow Bell). And the beans! Purple and green and blue-green, long or short, snaps and shellies and waxy ones, all waiting to be planted in the warm soil of my oh so slightly claybound Indiana garden. That first garden was sheer heaven.

After Southern Exposure, I found Mays Greenhouse in Bloomington, Indiana, where the three Mays sisters (Nancy, Helen, and Judy) and their father talked about gardens the way my mother did—with relish and gusto. I went out to look at their tomato starts, hoping to add some to my own slightly wimpy seedlings, and Mr. Mays was pottering around in the tomato shed. "Now then, young lady, do you favor a sweet tomato? I've got some Brandywines back here, and you'll love 'em, they're just like sugar." I told Mr. Mays that no, I tended to like a tomato that bit me back when I bit first, and he didn't argue a mite. He simply loaded me up with Heinz 1350, Rutgers, and Marglobe. "They've been around for a while, just like me, and I think these will be to your liking," he said, grinning up at me. "Now, if you need a sweet potato, I've got some dandy slips in right now. Step right this way. How do you favor 'em? Baked, boiled, or mashed?"

So I was hooked on heirlooms. It simply couldn't be helped. My mother and Grover Hughes led me to the precipice, and Dr. Jeff McCormack and the Mays family pushed me off.

I am by no means the world's greatest gardener. I'm not even in the bottom 10 percent, but I'll tell you one thing: I love to grow tomato seedlings better than almost anything else in this world. Putting those little seeds down in warm, moist seed starter, covering them up, and then turning the lights on for the first time is what comes to mind when I think of the word *hope*. The sight of a tomato seedling nosing up out of its covers is sheer joy. And the sight of pristine quarts of tomatoes put up against the chill of January and February produces in me a deep and glorious gloating. (Looking at those canned tomatoes nearly nine months after starting the seeds, I've often thought to myself that a human being can produce a baby in less time than it takes to make a quart of tomato sauce.)

I am by trade a librarian, not a gardener, and part of my job as a librarian is to design workshops that adults will find both useful and enjoyable. So in a brainstorming session with some of my coworkers in 2000, I announced that I would create a program to teach people how to start their own heirloom tomatoes, peppers, and eggplants from scratch. We would give the participants packets of seeds, show them how to plant the seeds properly and care for them, and send them off. I would also teach them how to transplant tomatoes using seedlings I had started four or five weeks earlier, and they would head home with some seedlings as well as their seeds.

In March of 2000, three people attended the program. We played in the dirt, talked about tomato recipes, and transplanted our little tomatoes, and every one of us went home happy. I thought I would never offer the program again—not for three people—but next spring the gardening bug bit, and in 2001 fifty people showed up. In 2004 I asked a few other libraries in other counties whether they would like me to come and do the program there, and even more people showed up. In 2006 the Appal Seeds Heirloom Seed Programs for Kentucky Public Libraries appeared at thirty-two libraries, the Kentucky Folk Art Center, and the University of Kentucky Arboretum. A record 1,923 participants showed up for this free program. They laughed, talked, ate, swapped big garden lies, and went home and grew tomatoes. I have done this every year since then.

The attendance goes up and down, and I still go to as many libraries as my schedule permits, which is right around fifteen in central Kentucky. This means that somewhere around 900 people show up at their public libraries every spring and leave about three hours later armed, somewhat haphazardly, with enough knowledge and raw materials to raise at least 200 tomato plants apiece, if they were to grow out all their seeds successfully. All the seeds are heirlooms, and they're all open-pollinated varieties, so these folks never have to buy another tomato seed in their lives if they don't want to.

Although I've had a few people tell me over the years that they've grown out every seed in their packets, I figure that if each person raises three plants, they each get about ninety pounds of tomatoes out of the program—or about $350 worth of fresh heirloom (and organic, if they grow it that way) produce. If they save their seeds, which we also teach them how to do via the fermentation method, these workshop participants wind up with $150 worth of seeds as well. That's not a bad feeling—to know that every year our public libraries are quietly working to put $450,000 worth of food on Kentuckians' tables.

In the meantime, we've saved several tomato varieties through the program, such as the Butler Skinner, a big red velvety beefsteak from Winchester. Depp's Pink Firefly, a large pink potato-leaf tomato that hails from Glasgow, came to us via archaeologist Cecil R. Ison, who received it as a gift from his friends and colleagues Dr. Fred Coy and his wife Emily Depp Coy. This variety has been offered and grown in the program many times, as has the Rose Beauty from Estill County, Maruskin's Andes from Winchester, and now Lena Mae Nolt's Holy Land, which hails from Casey County.

And in some counties the program means much, much more. It is a part of the people themselves, and I am happy to say that I am surrounded by others who are just as vegetable crazy as my family. In some counties—most notably Rowan County—participants bring bags of beans, buckets of Jerusalem artichokes, lettuce starts, and I don't know what-all (as

Part of an enthusiastic crowd gathered to hear Bill Best talk about seed saving at one of Julie Maruskin's library programs.

my grandmother would have said). And after the program is over, it's great to see everybody clutching their seeds and their plants and grinning like first-graders coming out of a candy store where everything is free.

In addition, I've been fortunate to meet the likes of Jere Gettle of Baker Creek Seeds, Dr. Bill Best of Sustainable Mountain Agriculture, Brook Elliott of the Appalachian Heirloom Seed Conservancy, tomato grower and enthusiast Darrell Merrell, and growers Merlyn and Mary Ann Niedens. The library program has been supported by my coworkers and colleagues for more than a decade, and I can say that there are no finer folks than those I have met across the state in dozens of public libraries. I thank each and every one of them for the success of our heirloom tomato crusade.

The final word? My hope is that long after my day, these programs and the plants they are built around will be passed on with all the joy that my great-great-grandparents passed on to a slender little girl with chestnut-brown hair and smoky blue eyes, who eventually passed that heritage of delight on to me.

Gary Perkins

Gary Perkins lives in Wayland, Kentucky, in Floyd County. In 2008 he retired as a supervisor with the Division of Mine Reclamation and Enforcement. He returned to work in 2010 as an environmental inspector with the Natural Resources and Environmental Protection Cabinet. In addition to gardening (fruits, vegetables, flowers, trees, and shrubs), his hobbies include genealogy, reading, Kentucky history, and photography. He has been involved with an interesting project undertaken by the Floyd County extension office to install small raised-bed gardens for low-income senior citizens. Here's Gary's story in his own words:

I'm not sure as to why I have collected or continue to collect and save seeds. It's something I have done for most of my life and will continue to do until I'm no longer able to garden. For me, it's sort of an addiction, and I feel that I would be missing or losing something very important should I no longer be saving seeds. I related several of the varieties to the people from whom they were collected. Many of these people were friends and family, and they gave me something they valued.

In the far eastern part of the state (coalfield section) I feel that there continues to be a decline in the number of gardens. In the summer there are areas where you can drive miles without seeing a garden. It's difficult for me to understand. The food pantries/banks continually have shortages, and children reportedly go hungry; however, fewer seem to be growing food. I do think that farther west, where the small tobacco farms were once located, there does appear to be an increase in market gardens and in home gardening.

Since early childhood I have had a fascination with growing and propagating various types of plants. I began collecting bean seeds more than thirty years ago, when I came to the sudden realization that some of the beans that our family had once planted no longer existed. Family members and neighbors that had kept and valued these seeds were no longer around. Up until that time it hadn't occurred to me, but it was very probable that only a small number of people, living within a radius of a few miles, had ever grown these particular varieties.

I was born and grew up in eastern Kentucky, near Hindman (Knott County). We lived along a small stream named Owens Branch, with neighbors of the same socioeconomic status. This area, as with most of southern Appalachia, had been stereotyped by the outside world as being poor and backward. My feeling was that they didn't know or understand the Appalachia that I knew. I feel fortunate to have grown up around such genuine and kind neighbors.

As late as the early 1960s, a good portion of the older generation in this locale were still living as they had before electricity, paved roads, and various modern conveniences had made their way into this part of Kentucky. These were folks that still dropped by for visits, obtained their water from hand-dug wells, kept milk cows, and plowed their gardens and fields with mules. These were people that had always relied upon the soil for sustenance. I feel that I have witnessed in my lifetime the loss of a large portion of our unique eastern Kentucky culture. In a small way, preserving some of the old bean varieties allows me to hold on to a little of my heritage.

I will have to admit that my initiation into the world of gardening was not by choice. Although my mother loved gardening, our vegetable garden was grown out of necessity. My father died at an early age, leaving my mother a widow in her mid-thirties. With several children to feed and a low-paying job, growing vegetables to supplement her small income was a logical choice. My earliest recollections are of tagging along with her in the garden.

Our garden was small compared to our neighbors; it may have been a half acre in size. We often utilized our neighbors' idle garden plots when they were available. The most important vegetable grown was the pole beans. Other basic vegetables we grew were potatoes, corn, tomatoes, broccoli, onions, lettuce, mustards, okra, turnips, cabbage, sweet potatoes, and peas. We grew the commercially available White Half-Runner beans and the Kentucky Wonders, but the most treasured were the greasy and cut-short varieties that had been obtained from family or neighbors.

In my mother's garden, sentiment did not factor into whether a particular variety was planted or discarded. Grandma's heirloom would be replaced if a variety came along that was more productive and had a superior flavor. Although we dried some of our beans for use as shucky beans, and some were pickled, the majority were canned, frozen, or used as fresh snap beans for cooking. We basically had one way of preparing our snap beans. After stringing and breaking up the beans, they were slowly cooked for several hours with a small amount of salted pork. All the beans we grew were judged solely on how well they tasted after being cooked in this manner.

Over the years I have obtained an estimated one hundred varieties of beans, with approximately half of them coming from the counties of Floyd, Knott, Letcher, and Morgan in the eastern portion of the state. Many were obtained from neighbors and family, and others obtained by seeking out gardeners and small farmers. I think I have enjoyed the process of collecting seeds as much as I have growing them. I felt as though I was setting out on an adventure when traveling to meet a fellow seed saver.

Many of the varieties that I've collected have descriptive names, such as the Buckeye Fall bean and the Greasy Grits; others were named for the location where they were obtained, such as the Perry County Cornfield; and still others were named for the person from whom the bean was obtained, such as the Irene Jones bean.

Below is a list of some of the varieties that I have collected, along with a short description. Some of my descriptions are brief, since I am working from memory and it has been several years since I have grown several of these varieties.

Red Striped Greasy: This is one of the varieties I remember my family growing during the 1960s. I had lost seed of this variety but in 2002 was able to regain seed from Gillis and Beulah Sturgill of Hindman. It had been ten years since they had grown this particular variety, but they still had seed stored in their freezer. They gave me a half pound of seed, and when planted, nearly all the seed germinated.

Brown Hastings: This variety was grown by my wife's great-grandmother, Leander Johnson, during the 1940s.

She had purchased the seed from a mining company's store located at Weeksbury, Kentucky. It has been passed down through the family for the past seventy years. It is possible that the company store obtained the seed from the H. G. Hasting Seed Company in Georgia.

Big John Bean: A white-seeded pole bean obtained during the 1990s in the Dry Fork area of Letcher County. A listing in the *Seed Savers 2004 Yearbook* states the following: "has been in Billy Adams' family for over 200 years, brought to Kentucky by Big John Combs around the time of the Revolutionary War." I don't know how valid this story is, but I believe this is the same John Combs that was my great-great-great-great-great-great-grandfather. Family historians state that he came to Kentucky with his eight sons shortly after the Revolutionary War. The Combs family had been in Shenandoah County, Virginia; Surry County, North Carolina; and Sullivan County, Tennessee, prior to finding a permanent home in eastern Kentucky.

John's Creek Junk Pile: A gray-seeded pole bean, with vines eight to ten feet. During the 1990s, while working in the John's Creek area of Pike County as an environmental inspector, I happened on some garbage that had been dumped in an area that I was in charge of overseeing. While looking through some of the papers for an address of the guilty party, I happened upon a small bag of bean seed. The following spring, several of the seed germinated.

Sugar Bean: A white-seeded pole bean collected in Letcher County. A number of years ago, I corresponded with Ms. Susan Mosley of Pine Bluff, Arkansas, via a website. She stated that she had a relative located in Virginia who had collected this bean several years ago while working as a salesman in Letcher and Perry Counties of Kentucky.

White Fall Bean: A white oval-shaped bean collected from Ivan Johnson of Right Beaver Creek in Knott County.

Mountain Climber: A six- to eight-foot pole bean obtained from Letcher County. The seeds are light brown, similar to the color of the Kentucky Goose bean. Seeds have a tendency to sprout in the pod while still on the vine.

Bailey's Six Weeks Bean: Obtained from Sam Bailey of Lawrence County, Kentucky. Mr. Bailey estimated that this bean had been in his family for 150 years. It is similar to the pink half-runner bean sold in these parts by Southern States Cooperative. Vines grow to two to three feet and do best with some type of support.

Brimstone: When completely dried, the beans are a tan color. However, when shelled out in the green shell stage and viewed from a distance, the beans have the appearance of glowing embers. Obtained from Letcher County.

Coon Bean (Raccoon Bean): This bean got the name "Coon" because the seed color resembles that of the North American raccoon. It is a large, vining pole bean, with pods six to eight inches. The seeds are gray and brown colored, with some frosting. This bean was obtained from my mother-in-law, Myrtle Bates of Wayland, and has been grown in her family for more than seventy years. Garden writer John Yeoman has reported that the pods of this bean exceed ten inches when grown in his garden in the British Isles. Originally obtained from Knott County.

Buckeye Fall Bean #1: A six- to eight-foot pole bean; the seeds are oval shaped and are two-thirds white and one-third maroon in color. My wife's uncle, John Bates of Dema, Knott County, received these seeds in a trade with a coworker.

Buckeye Fall Bean #2: Obtained from Minnie Smith in Morgan County, Kentucky. This bean has a slightly different color variation than the above-mentioned Buckeye Fall Bean #1.

Dry Fork Greasy: A six- to eight-foot pole bean obtained in Letcher County.

White Hastings: This bean, collected in the Dry Fork section of Letcher County, is a small white-seeded pole bean with vines growing six to eight feet. My mother-in-law's family grew a bean by the same name in the 1950s.

White Seeded Kentucky Wonder: A pole bean obtained from Morgan County.

Red Fall Bean: A six- to eight-foot pole bean obtained from Knott County.

Sulphur Bean: A bunch (bush-type) bean with oval-shaped seeds; the seeds are a light brown (sulphur) color. Our family grew this bean during the 1960s. The original seeds were obtained from a local retail source.

Perry County Cornfield: Large white-seeded pole bean obtained from a roadside vendor near Hazard (Perry County) in the 1980s.

Creasyback: A large cornfield-type pole bean collected in Letcher County. The seed is oddly shaped, with a curved crease on one side. (Note: This is a different bean than the Striped Creaseback, aka Genuine Cornfield.)

Brown Greasy (Speckled Greasy): A prolific greasy-type bean with vines six to eight feet (from Knott County).

Striped Bunch Bean #1: A three- to five-foot half-runner-type pole bean collected from Alvin and Delia "Dilly" Calhoun of Hindman.

Striped Bunch Bean #2: From Myrtle Bates of Wayland.

Striped Bunch Bean #3: Obtained in Morgan County.

Myrtle's White Greasy: A white-seeded greasy-type pole bean obtained from Myrtle Bates, who found it in Knott County.

Late Speckled Greasy: Late-maturing pole bean with vines six to eight feet (from Knott County).

White Half-Runner: My original seed was purchased from a local hardware store. Over the years, I have carefully selected the seeds I saved, being careful to avoid any from the tough-hulled plants.

Cream Colored Bunch Bean: Bush-type bean obtained in Morgan County.

Harold (Conley) Bean: A six- to eight-foot pole bean obtained from longtime market grower Gillis Sturgill of Hindman. The seeds are tan in color, with small brown stripes. Mr. Sturgill described it as an excellent snap and soup bean.

Goose Bean (Kentucky): A six- to ten-foot pole bean with vigorous vines that need strong trellising. This is one of the beans my family and neighbors grew during the 1960s. The seed of the Goose Bean is large and tannish brown in color. My seed was obtained in Knott County.

Hook Bean: The bean pods are shaped like a hook. Obtained from Myrtle Bates of Wayland (Floyd County).

Irene Jones Bean: A six- to eight-foot pole bean recently obtained from Irene Jones of Arnold's Fork in Knott County.

Speckled Bunch Bean: Bush-type bean with small runners obtained in Morgan County.

Ivis White Greasy: White-seeded greasy-type bean obtained near Hindman.

Cornfield Greasy: White-seeded greasy-type bean obtained in Morgan County.

Tender Fall Bean: Cranberry-type fall bean obtained in Morgan County.

German Cornfield Bean: A large pole bean with seeds similar to the color of the Turkey Craw. Obtained near Mayking in Letcher County.

Old Fashion Cornfield Peas: Cowpeas obtained from Jimmy and Minnie Smith of Stingy Creek Road, Morgan County. Mr. Smith remembers his family (and other families in the same area) growing these when he was a child.

Myrtle's Cowpeas: I obtained these from my mother-in-law, Myrtle Bates of Wayland. Her family has saved this seed for more than seventy years, and she can remember her grandmother cooking these cowpeas for her more than sixty years ago.

Littlee White Greasy Cut-Short: This bean has been in my family for as long as I can remember. It may have been originally obtained in Perry County.

Coffee Bean: Seed obtained from Ivan Johnson of Right Beaver Creek in Knott County.

Colored Greasy Grit: Obtained in Morgan County.

White Greasy Grit: Obtained in Morgan County.

Roger H. Postley

A retired science teacher, Roger Postley gardens in his backyard in the suburbs of Lexington. He sells tomatoes and shares his recipes at the Bluegrass Farmers' Market. He is especially well known for raising and selling transplants of heirlooms, helping lots of gardeners get a start.

Though I grew up in the New York City suburbs and, after high school, moved to a dairy farm in upstate New York, I have lived in Lexington, Kentucky, since graduate school in 1967. With the exception of some college years, I have always had a vegetable garden of some nature. I always raised some tomatoes! My dad started me in gardening at about age five, but I didn't do much on my own until the 1970s.

In the early 1990s I found articles about "heirloom tomatoes" that "actually tasted great" and decided I had to try some! Even today, it is difficult to find many varieties of heirloom tomato plants, but back then, it was nearly impossible. I had taught Russ Madison's son and daughter (Proper Plants, Lexington, Kentucky) and approached him about raising transplants for me if I bought the seeds. I would get two or three plants of each variety, and the rest were his. I "blame" Russ for getting me into the heirloom transplant "business." This eventually led to a large garden (predominantly tomatoes and peppers), heirloom seed saving, and raising large numbers of transplants for sale at the local Bluegrass Farmers' Market. In 2013 I grew out 67 varieties of tomatoes and 9 varieties of peppers and started about 1,400 transplants.

To clarify matters, I need to explain the distinctions between hybrids, heirlooms, and OP (open-pollinated) plants. Hybrids are crosses between different varieties that were intentionally bred to ensure that certain shape, taste, or color characteristics are present in the fruits they produce. There is nothing inherently bad about hybrids, but you cannot save their seed, as the seed will NOT produce the same fruit as the parent plant. With only one or two exceptions, I do not raise any hybrids and must purchase new seed every year for those I do grow.

All heirloom plants must be open pollinated, but not all OP plants are heirlooms. It is generally accepted that to be an heirloom, a variety must have been in continuous production for forty to fifty years! OP plants, barring spontaneous "sports" or mutations, yield seeds that produce plants identical to the parent plant. The majority of what I raise are heirlooms, with a smattering of non-heirloom OPs and a few hybrids. A considerable number of the tomatoes I raise are Kentucky or Appalachian heirlooms.

I love heirloom tomatoes for two main reasons: taste and variety. My signature line on e-mails is, "I never met a tomato I didn't like—then I went to a grocery!" This is a sad commentary on the palatability and mouth-feel of commercial tomatoes developed primarily for ease of machine picking, shelf life, and "durability"—NOT taste! With heirlooms, you can raise tomatoes that have almost any range of softness or firmness you desire. They have different uses, including cooking, slicing, salads, salsas, and sauces. They come in sweet, balanced, tart (there is actually no such thing as a low- or high-acid tomato—they all have about the same pH, but they vary greatly in sugar content), tangy, juicy, or dry.

Heirloom tomatoes come in an amazing variety of shapes, sizes, and colors, but the one characteristic they all seem to have is great flavor. Why else would a variety be grown for forty or fifty years? Heirloom tomatoes are available in shades of white (usually a very pale tan or yellow), pink (these tend to be sweeter), obviously red (but don't limit yourself to this color), purple (rich flavor), brown (knock-your-socks-off flavor), black (extremely flavorful and actually more like the color of a bad bruise), yellow (rich and juicy), orange (full flavor), striped (many are OP but not heirloom), and bicolor (blotched or streaked in red and orange/yellow). They range in size from one-quarter inch (delicious, but a pain to pick) to salad size to three-pound monsters (a BLT waiting to happen). They can be spherical, lobed, flattened, and rounded (beefsteak), heart-shaped, pleated, or elongated (shaped like a fat carrot).

A majority of heirloom tomatoes are indeterminate plants. This means that they produce fruits continuously, from the initial bearing until they are killed by disease or frost. Most tend to be tall—frequently over six feet. (Determinate plants are almost always short and usually produce one major "flush" of fruits and then die.) This means that heirloom tomato plants require tall support. Most commercial tomato cages are too flimsy and too short. There are numerous articles online (including my own) on how to build your own strong, tall cages from concrete remesh (reinforcing wire). Most of the ones I build are eighteen inches in diameter and six and a half feet tall. My plants typically grow out the top by one or two feet! They may have either regular leaves or potato leaves.

An obvious progression from raising heirloom tomatoes is saving their seeds so that you can keep a variety around without having to purchase new seeds each year—if that variety is still available. Most tomato seed savers use personal variations on the same tried-and-true method: fermentation. Although you can just squeeze the seeds out of a tomato onto a plate or paper towel and air-dry them, this does not remove the gel (a germination inhibitor) or the seed-borne bacteria, viruses, and fungi.

To save seeds via the fermentation method, I use twelve-ounce clear plastic cups. Label each cup with the variety name and the date. Cut the tomato in half (or smaller pieces, if necessary) and squeeze or scrape the seeds and gel into the cup. Don't worry if some of the tomato pulp falls in, but try to remove the bigger pieces. Crush and mix everything with your fingers or with a fork or spoon. Fill the cup with water. I set the cups on outdoor windowsills that are protected by the eaves from rainfall. I let the seeds ferment for four to six days, stirring each cup several times a day. You will get a layer of mold on the surface—this is actually what you want! The cups may have a strong odor (hence the reason for fermenting them outside), and they will attract many small insects—but don't worry! At the end of the fermenting period, stir the cup and let the seeds settle. Nonviable seeds and most of the pulp and mold will float on the surface. Carefully pour this off.

Next, move the cups to a sink and fill them with water. Let the seeds settle, and pour off the water and remaining solids. Repeat until the water is clear and the seeds are clean. I then pour the seeds with a small amount of water onto a coated paper plate. Carefully pour off as much of the water as possible. After the first day, stir the seeds so they don't stick to the plate, and spread them out. Let the seeds dry for one to three weeks at room temperature in a low-humidity room. I store my seeds in small, labeled zip-type plastic bags. If kept at room temperature and in the dark, the seeds will remain viable for at least five years! (Commercial seed companies don't tell you this, as they want you to buy new seeds every year.)

For the past several years, I have concentrated more and more on raising Kentucky and Appalachian heirloom tomatoes—I have literally hundreds of varieties to choose from! Again, most of these varieties won't be available as transplants, so you'll have to learn how to start your own plants from your own saved seed, which is an obvious extension of your home gardening. The majority of the heirloom varieties I grow originated from swapped, gifted, or purchased seed. I keep the varieties going with my own saved seeds.

Just as a teaser, here are the Kentucky and Appalachian varieties I anticipated having for 2014:

Australian Heart (Thieneman family): Brought to Louisville by a returning World War II navy veteran; produces a medium to large red tomato that is heart-shaped, extremely tasty, and very juicy.

Barnes Mountain Yellow: Estill County heirloom; produces one-pound yellow, juicy beefsteak fruits with very good flavor.

Black Mountain Pink: Harlan County heirloom; produces one-pound (medium-large) pink, juicy beefsteak fruits with good flavor.

Buckeye Yellow: Madison County heirloom; produces one- to two-pound yellow beefsteaks with red blotches and a mild, sweet flavor.

Butler Skinner: Clark County family heirloom. The seeds, plants, and fruits were once given as gifts during political campaigns. The twelve-ounce round, pink fruits have a full-balanced, sweet flavor.

David's Pink: Grayson County heirloom; this six- to twelve-ounce beefsteak is very meaty and extremely tasty.

Depp's Pink Firefly: Depp's family heirloom from Glasgow produces numerous large, flattened beefsteak fruits that are pink with light spots and have a fantastic flavor.

El Ifino: An orange variety discovered by me! Named from the old joke: what do you get when you cross an elephant with a rhinoceros? Answer: El Ifino, a sport of Lillian's Yellow Heirloom that produces a medium to large orange beefsteak with a great juicy full flavor. Also available as El Ifino–Red.

Frank's Large Red: An eastern Kentucky heirloom that produces huge (one and a half to two pounds) red, flattened, slightly lobed fruits that are very juicy and have a fantastic flavor.

Granny Cantrell's German Red: A West Liberty heir-

Roger Postley trading heirloom seeds at a Sustainable Mountain Agriculture Center event.

loom raised since 1945 by Lettie Cantrell, who passed away in 2005. This is a fantastic, very large red beefsteak with outstanding juicy flavor.

Hazelfield Farm: This OP variety from Wheatley, Kentucky, isn't old enough to be an heirloom. It produces four- to eight-ounce flattened red beefsteaks with slightly ribbed shoulders; they are juicy, have a very good full tangy tomato flavor, and are great for cooking.

Kennard's Big Red: This Johnson County heirloom is actually a dark pink beefsteak with an outstanding sweet, tangy taste and very few seeds. The flesh is extremely dense but still juicy! Fruits weigh a pound or more.

Kentucky Beefsteak: Eastern Kentucky heirloom. This large, yellow, flattened beefsteak is a late but heavy producer with an outstanding flavor.

Lumpy Red: Corbin, Kentucky, family heirloom that is highly productive and yields heavily fluted medium to large red fruits that are very tasty.

Max's Large Green: Family heirloom from Hardin County with firm, juicy, bright green flesh in a sixteen- to twenty-ounce flattened beefsteak. The skin has an amber tinge when ripe. It has a delicious, tangy full flavor and makes a lovely sandwich slicer. Another BLT just waiting to happen!

Minnie's Pinstripe: This Bowling Green, Kentucky, heirloom is a one-pound bicolor beefsteak that is very juicy and sweet.

Monk: A family heirloom from Nicholasville that produces slightly irregular, very large, and extremely tasty beefsteak fruits on large, strong vines.

Old Kentucky: A large yellow southeastern Kentucky heirloom with an even better flavor than the Kentucky Beefsteak!

Purple Dog Creek: Hart County heirloom with huge, meaty, purple-red fruits and excellent flavor.

Rose Beauty: From Estill and Jackson Counties, the Rose family heirloom beefsteak produces pale yellow large to extra large fruits (some have pink streaks). The robust flavor is similar to large red beefsteak varieties. Very juicy.

Super Choice: A Liberty, Kentucky, Amish family heirloom that produces numerous very large, beet-red, flattened fruits with a fantastic flavor.

Uncle Mark Bagby: This western Kentucky family heirloom has very large pink fruits and excellent flavor.

Vinson Watts: Large (one to two pounds) flattened pink fruit with excellent flavor. This old heirloom was grown by the late Vinson Watts in the Morehead area but originated in Lee County, Virginia. It is juicy and full of flavor and great on sandwiches!

In addition to these Kentucky family heirloom tomatoes, I raise heirloom tomatoes that originated in Florida, Georgia, Missouri, North Carolina, Ohio, Tennessee, Utah, Virginia, and West Virginia. They are joined by heirloom tomatoes originally from Australia, Burkina Faso, Canada, England, France, Germany, Hungary, Italy, Lebanon, Mexico, New Zealand, Portugal, Russia, and Ukraine. There are heirloom tomatoes from countries all over the world. You can find heirloom seeds that will produce fruits with almost any shape, color, or taste characteristics!

So jump into the heirloom pool—the water is fine. And all aspects of gardening, seed saving, and growing your own transplants are fun!

Susana Lein

Susana Lein has developed a self-sustaining market farm from degraded land without topsoil on the side of a mountain—where many said it couldn't be done. Susana, who comes from a Midwest farming background, started Salamander Springs Farm near Berea, Kentucky, in 2000. She cleared an overgrown meadow and built a gravity-fed spring water system. She camped on the land for most of the year after building an open-air kitchen shack with posts of cleared locust and cedar trees, pallets, salvage, and slab wood—from the lumber milled for the off-grid, passive solar house where she now lives.

Without tillage and using few purchased inputs, Salamander Springs Farm produces grains, dry beans, produce, fruits, flowers, and herbs for local farmers' markets, community-supported agriculture shares, an online store, and retail outlets. Susana teaches permaculture, demonstrating her farm's systems for producing sta-

Susana Lein with workshop participants in the cornfield at her Salamander
Springs Farm. (Susana Lein/Salamander Springs Farm)

ple grains and dry beans, inspired by Japanese rice farmer Masu-
nobu Fukuoka's seminal work, and for producing no-till cornmeal
and popcorn using "Three Sisters" Native American practices.

Susana selected, bred, and developed her popular rainbow-
colored corn from seed passed down through the generations by

Daymon Morgan's family in Leslie County. This is another perfect example of how a variety can develop and change over the years when seed is saved and selected for size, color, vigor, and drought tolerance, as Susana explains:

> Selecting seed from locally adapted heirloom crops has helped sustain Salamander Springs Farm over the years. Local heirloom varieties produce dependably without outside inputs. Often the same-named varieties from heirloom seed catalogs lack the vigor, productivity, and flavor of my locally selected seed (such as tomatoes, salad greens, peppers, beans, and corn). At the market, both old-timers and younger folks appreciate the rich flavor and nutrient density of these heirlooms.
>
> To truly sustain ourselves in an uncertain future, farmers need to be saving more of the seed of what we grow. Having to purchase seed puts us at the mercy of a fallible system. I use predominantly open-pollinated varieties for this reason. An open-pollinated vegetable variety, when saved over the generations, becomes an heirloom. Even vegetable varieties that cross readily, like carrots or brassicas, can be saved if isolated in time. I save the seed of winter varieties when they go to seed in the spring.

The late Daymon Morgan was such an important part of my first year in eastern Kentucky and will always be part of Salamander Springs Farm as I plant, harvest, and mill the rainbow-colored corn I've developed from seed I got from his corn.

I remember Daymon's love and understanding of the Appalachian forest when I hike my own. Much of my knowledge of forest medicinals started with him.

Salamander Springs Farm apprentice Dori Stone picking heirloom cornfield beans growing with Kentucky Rainbow corn—the "Three Sisters" Native American way. (Susana Lein/Salamander Springs Farm)

Daymon was an open-minded and curious soul who valued the wisdom of many cultures, not just his own. Some of my favorite memories are helping him practice his Spanish acquired from his trip to Nicaragua.

In 1999 Daymon gifted me seed of a dent corn grown for generations in his family in Leslie County, Kentucky. Though he still called it Bloody Butcher, it had crossed over the years to have a diverse parentage and few red ears. A

Daymon Morgan of Leslie County on his front porch looking out over his last garden. (Dobree Adams)

majority of cobs were red, but the kernels were predominantly white with some blue mixed in. The gene pool occasionally expressed Bloody Butcher in entirely red-kerneled ears but most often represented the white-kerneled Tennessee Red Cob (or possibly Neal's Paymaster), Hickory King (with fewer rows on a thinner white cob), and Blue Clarage. Also represented was a thin-kerneled gourdseed-type corn with ears of painted orange kernels (my favorite and now more prevalent).

About 35 to 40 percent of the corn gene pool now has red cobs. After seventeen years of selecting and breeding, ear size has increased dramatically, along with the array of kernel colors and tighter husk closure (for ear protection). Primary ears with more than fourteen rows of kernels have increased from much less than half to 80 to 90 percent. The vigor of this developed crossbreed is astounding to those who see it in the field. Most plants now have two ears and measure more than sixteen feet tall.

In 2008 the Virginia-based Southern Exposure Seed Exchange (SESE) approached me after my workshop at the Carolina Farm Stewardship Association conference in South Carolina. I began selling seed for its catalog, and I originally named it Daymon Morgan's Kentucky Butcher in an attempt to include some of its history. I have since changed this long name to Kentucky Rainbow to reflect its diverse parentage and my selective breeding changes.

Kentucky Rainbow has become an immensely productive, drought-tolerant corn in my no-till "Three Sisters" system with squash and beans. Customers across the country rave about its incredibly flavorful cornmeal. The sweet roasting ears remind me of my time in Central and South America. The core of my knowledge of corn seed selection and development came from the years I worked with Mayan *Pokomchi* farmers of Alta Verapaz, especially the late Don Gavino Ca'al, from whom I learned so much. My more recent research into how genes represent themselves in corn reproduction has substantiated the wisdom of their methods. Each hamlet of the Pokomchi people where I lived during the 1990s was identified by the color of its corn—blue, white, red, and all shades in between. Seed was selected for size, vigor, and tolerance of difficult conditions without irrigation or added amendments. Because of the work of these Central American farmers over the past 500 years, fourteen-

Daymon Morgan's last garden and cornfield, summer of 2014. (Dobree Adams)

inch ears of corn were developed from *teocinte*, the parent of corn, which was the size of a grain of wheat.

While Kentucky's corn production was reduced more than 50 percent by the intense drought of 2012, tour groups at Salamander Springs Farm were amazed to find a healthy, productive cornfield without irrigation. Ears can reach twelve inches long and yield a pound of corn each. Cornstalks average twelve to fourteen feet and have reached as tall as eighteen and a half feet!

Because of its overwhelming popularity in the catalog, SESE now contracts several other seed growers in the southeast who use foundation seed stock from Salamander Springs Farm. When the seeds of so many heirloom corn varieties are showing GMO (genetically modified organism) contamination from wind-borne pollen, my location

has been a blessing—on an Appalachian ridgetop surrounded by forest and a great distance from any conven-
tional cornfield.

"How many feet in a meter?" In disbelief, Franzi Habith, a farm apprentice
from Austria, measures cornstalks at Salamander Springs Farm. (Susana
Lein/Salamander Springs Farm)

In 2011 Garland Elkins, a neighbor in Rockcastle County, Kentucky, gifted me seed of several heirloom varieties of pole (cornfield) beans when he delivered some geese. Some were varieties I already grew (like White Greasy and Goose beans), but a 200-year-old heirloom bean he called the Elkins bean had a special family history. Garland's great-grandmother, Robinette, was eight years old when the family journeyed over the mountains from North Carolina to Kentucky sometime before the Civil War. She was charged with keeping the seeds of this special bean safe and dry in her apron pocket. Later, she often told the story of how the family had rested on their journey at a beautiful place with fresh springwater where Virginia, West Virginia, and Kentucky meet. When Garland was a teenager in the 1950s, his family took his elderly grandfather to see this place—in present-day Breaks National Park near Elkhorn, Kentucky.

The Elkins bean, now a treasured part of the cornfield at Salamander Springs, looks somewhat like the popular half-runner bean, but with an unsurpassed hearty flavor. It is delicious as a fresh string bean in the summer and as a protein staple in the fall and winter. Garland's family shucked the later yellowing beans for "the best soup bean there is." He still enjoys many staples of our ancestors, who, by necessity, lived entirely from the land—soup beans, squash, cornbread, grits, and crackling bread made with the fat of a butchered hog. Now in his seventies, Garland Elkins wants his treasured bean seeds to live on. May we keep them growing and feeding the people in our community!

The High Price of Cheap Food

A Case for Heirlooms

While watching a television news program recently, I was reminded of the rapidly rising rate of obesity in this country. Numerous doctors and health organizations are now sounding the alarm about the growing costs of medical care at all levels from childhood upward. For some years, children have been developing diabetes, and heart attacks are not uncommon in younger age groups. Many health professionals are now predicting that the current younger generation will be the first to have a shorter life span than their parents' generation.

As someone who holds degrees in physical education and health and has coached sports for twenty-five years, I am dismayed that so many schools no longer require physical education or are doing away with physical education programs entirely. Colleges are even worse. Many of them have no physical education requirement at all, the programs having been voted out by the least physically fit faculty members in many instances.

School lunch programs no longer focus on healthy food

Passing it on! Our great-grandchildren Greta, Lilah, and Peter Hess learning to string beans, August 2015.

choices, and many students go all day without eating a fruit or a vegetable, fresh or cooked, either at school or at home. Sweeteners made from corn show up in almost all prepared foods sold in groceries—everything from candies to breads to drinks and many types of canned and frozen foods. I was stunned by the high percentage of corn in our bodies, as shown by numerous studies.

We have to salute First Lady Michelle Obama for focusing our attention on changing all this with her White House kitchen garden, her work on improving school lunch programs, and her emphasis on the importance of moving our bodies and living an active life.

My grandson Brian, who is now in his twenties and in college, has worked for me during the summers since he was twelve years old. I always let him choose where we stop to eat on our way home from the Lexington Farmers' Market, and we

Working in the beans. My great helpers: grandsons Brian Best and Nolan Toti.

usually end up at a fast-food restaurant that offers both healthy and not-so-healthy choices. Coming back from Lexington one Saturday a few years ago, I ordered a sandwich that included lettuce, tomato, and slices of onion. I opened up my sandwich and found what looked like the white heirloom tomatoes I had been selling just an hour earlier at the farmers' market. The only problem: the tomato on my sandwich was supposed to be red. If my sandwich had included a slice of vine-ripened tomato, I would have gained vitamins A and C as well as lycopene and other vitamins and minerals. Given the sad state of my anemic tomato slice, at best it provided me with a small amount of water and some fiber and carbohydrate. Certainly it added no flavor to my sandwich. My grandson removed his tomato slice, but being of the old school, where I had been taught to eat everything on my plate, I suffered through mine and confirmed the lesson that modern foods are lacking in flavor, texture, and nutrition.

While watching television programs with historical themes, I am always stunned by how modern mechanization has changed the labor situation in agriculture and the nutrient quality of fruits and vegetables. One person running a half-million-dollar machine can now perform the work previously done by hundreds of humans. Plants have been designed to accommodate these machines; fruits and vegetables all ripen at the same time to facilitate the harvesting process. It seems that the primary reason for all this is to fatten the bottom line of the large farm corporations. I have never heard commentators speak about improved nutrition, and they obviously don't admit to lowering nutritional standards by producing cheap foods.

Machines have been developed to harvest rice, wheat, barley, oats, and other grains with incredible speed. Tomato-picking machines scoop up the entire plant, shake the hard tomatoes loose, and then spit the vines back onto the soil. Human beings enter into the process along the automated assembly line as they sort out the worst of the tomatoes. The goal is to end up with a pile of tomatoes of a uniform size and ripeness—or lack thereof.

Bean-picking machines can harvest designer beans with their designer genes very swiftly. The vines have been designed to thrust their beans several inches off the ground to accommodate mechanical harvesting. The beans are straight and of a uniform size, and they can be picked all at once, without having to worry about successive bloom sets. Thousands of bean varieties have been reduced to twenty or so that are well suited to mechanical harvest. All of them are remarkably similar in shape and color and have plenty of toughness bred in to prevent breakage during harvest. One automotive company advertises its trucks as being tough, but seed companies never advertise their beans as being tough. But beans, like trucks, are now tough. We are advised to "Pick while young and tender" and "Do not let lumps appear in your beans." The lumps, of course, are the bean seeds. One company proudly advertised its beans as being "grass-like," but who wants to eat a grass-like bean? Even organic seed companies have bought into the new definition of the bean and promote the same varieties as other seed companies.

Nut trees have been designed with a tough collar a few feet up so that they can be hugged and shaken by a machine that causes the nuts to drop to the ground, where they are swept into windrows and then scooped up by another machine. We are now developing machines that can shake blueberries and raspberries off bushes and apples off trees.

Plant Breeding for the New Agriculture

Back in the early 1980s, a severe hailstorm convinced me that I would be better off growing mostly heirloom tomatoes and selling them at farmers' markets and to restaurants. The hail damage meant that few if any of my tomatoes would be high-grade enough for the packing plant to accept. However, once the tomatoes were completely ripe, farmers' market customers were quite happy with them, notwithstanding the small hail scars. They wanted good-tasting tomatoes, and I had plenty of them. Four years ago, I stopped by a large tomato field in an adjoining state to see for myself how much tomato growing and harvesting had changed since 1984—the last time I sold commercial tomatoes to a packing plant. My visit was a revelation.

As I stood at the edge of the very large tomato field, I observed dozens of migrant workers picking green tomatoes and dropping them into white buckets that probably held at least five gallons of tomatoes. Then they poured the tomatoes into larger twenty-bushel containers on a flatbed truck. They worked quickly, since they were being paid by the quantity picked and not by the hour. There was no worry about bruising the green tomatoes.

The supervisor, also a migrant worker, was standing by his pickup truck, answering occasional questions from the field hands and generally making sure the operation ran smoothly. I noticed three twenty-five-pound boxes of ripe tomatoes, one of them open, in the bed of his pickup and asked if I might buy three of those tomatoes. He seemed pleased that I wanted to try some and replied, "No, but I will give them to you." Since all the tomatoes being picked were green (the workers had been instructed not to pick the ripe ones), I wondered where the three boxes of ripe tomatoes had come from. The supervisor explained that the pickers occasionally missed some of the tomatoes on the bottoms of the vines, and they ripened. He would then walk through the field, pick the few ripe tomatoes he found, and sell them to individuals who happened by.

It was unlikely that any grocery chain customer or fast-food patron would eat a vine-ripened tomato from that field, and I wondered what such a tomato might taste like. So later that night, I decided to try one. I ate it raw with just a little salt, as was my practice when eating my own freshly picked vine-ripened tomatoes, just to evaluate the taste of the tomato (I didn't want the ingredients of a sandwich to interfere with the taste test). I was shocked at the very bland taste and watery texture of the tomato. It was neither acid nor sweet tasting and had no tomato smell. Yet it was completely ripe and even somewhat tender to the touch. I took the two remaining tomatoes to the Lexington Farmers' Market two days later, cut them up, and let some of the other vendors taste them. Their reaction was the same as mine—genuine shock that a recently picked vine-ripened tomato was so watery and had virtually no flavor.

That same year I stopped to buy some tomatoes at a stand in

a county famous for its vine-ripened high-tunnel tomatoes. I was heading to North Carolina for our family reunion in late May. I am always expected to bring the tomatoes. Sometimes I have enough of my own high-tunnel tomatoes to take, but that year a had only a few. In the past I had found that the tomatoes sold at this stand tasted similar to mine, but to my surprise, as I sliced the tomatoes and set them out on the picnic tables, I discovered that they had very little flavor. Later that summer I went back to that same stand and asked about the difference. The vendor told me, with a perfectly straight face, that they had gone to a higher-yielding variety and didn't expect their customers to notice any difference in flavor and overall quality. I had certainly noticed the difference, as had the other tomato lovers in my extended family. And I was certain the grower had noticed as well.

I happen to know that tomato growers in that area are well aware of differences in the quality of different varieties of tomatoes. About four summers prior to the experience I just described, I stopped by that same stand to compare their toma-toes to my own and also to check out the prices they were get-ting. It was the last week of July, and tomato prices had already gone into their midsummer slump. Red-colored commercial tomatoes on the front porch of the packing shed were going for $10 per bushel, a very low price for good tomatoes. In contrast, heirloom Yellow German tomatoes, immediately adjacent to the commercial ones, were going for $80 per bushel and were sell-ing quickly. The Yellow German, of which there are many vari-ants, produces large yellow fruit with red streaks and blotches throughout when fully ripe. It is the sweetest of the most com-monly produced heirloom tomatoes. Like most other heirlooms, its shelf life is much shorter than tomatoes grown for shipping. But that doesn't stop customers from paying a premium price to get what they like.

The Games Seed Companies Play

At one time, seed companies were largely family operations that did the plant breeding themselves. Back then, it was easy to find quality fruit and vegetable seeds. Early seed companies sought out the best among family and community fruits and vegetables, and truth was the norm when it came to describing what one could expect to get when placing an order.

However, once seed companies started cannibalizing one another and feed/fuel/seed/fertilizer/chemical conglomerates began purchasing seed companies, each new offering had to be taken with a grain of salt. Worse yet, even the older varieties had to be viewed with a jaundiced eye. The best case in point is the White Half-Runner bean.

For decades, the White Half-Runner was one of the favorite beans grown in the southern Appalachians. It was good for eating fresh, for canning, and for making shuck beans, and in many areas it was almost the only bean grown. Some of the major seed companies took notice and began contracting with commercial seed growers to supply them with large amounts of cheap seeds that could also be sold cheaply. Things went well for a number of years, and fewer and fewer gardeners went to the trouble of saving their own half-runner seeds. But gardeners gradually noticed that a few beans were tougher than the rest. Some beans were flat, and the seeds appeared very slowly. The flat beans were so tough that it was almost impossible to string and break them for cooking purposes. Worse yet, the toughness persisted through the cooking process, leaving many gardeners wondering whether they had forgotten how to cook beans. Some beans in a pot appeared to be well done, while others seemed to need a lot more cooking time. And yet no matter how long the beans were cooked, the tough ones remained tough. The beans were clearly of two (and probably more) different types.

When farm supply store owners complained to the growers, usually located 2,000 miles away, they were told that the seeds were as good as they had always been. But in reality, there was virtually no quality control. With the seeds being grown in fields of 1,000-plus acres, no one was taking the time to go through the fields prior to harvest and pull up the vines of tough beans, which can be identified by their different leaf structures. As each season produced a new crop of bean seeds that included the rogue beans, enough crossing took place to increase the percentage of tough beans, as the tough bean genes were apparently dominant. Finally, by about the summer of 2006, tough beans became the majority. One lady at the Lexington Farmers' Market who regularly bought twelve bushels of half-runners for canning told me she had to throw away seven of those bushels. She had gotten accustomed to throwing away nearly half the beans she purchased, but when she had to throw away more than half of them, that was the straw that broke the camel's back for her. She started buying bushels of greasy beans from me instead and had to throw away none of them. She could pay twice the price per pound for greasy beans, get the same amount for canning, and end up with a superior product in terms of both flavor and tenderness.

Her experiences have been borne out by many other farmers' market customers. And yet farm and hardware stores continue to buy and sell the same contaminated seeds, claiming they have no other source of half-runner seeds. Most do so reluctantly, but there is still some demand for half-runner seeds, poor quality notwithstanding. The commercial growers of the contaminated seeds claim there is no problem: it's all in the consumers' minds. But I'm sure they know better.

This was all brought home to me when farm store owners and operators in three states contacted me and wanted to buy my entire stock of heirloom bean seeds, especially the heirloom half-runners I had started selling a few years earlier. When I asked why, the story was always the same: Customers were refusing to buy the seeds that produced tough beans. They were just saying no. Customers are now buying heirloom half-runners in droves from my website.

A Flawed Business Model

It is well known that there are far fewer farms and farmers in the United States than there used to be, and the farms that remain are becoming larger. Yet an opposite movement is also taking place: many small farms, ranging from half an acre to several acres, are springing up. These smaller farms are marketing their produce at farmers' markets and directly to chefs and specialty food markets, where people are willing to pay more for quality food. Other small farms are becoming involved in community-supported agriculture (CSA), whereby customers purchase a share in the farm's produce for a given period of time and, in return, receive a specified amount of fruits and vegetables on an agreed-upon schedule. These customers assume a certain degree of risk, in the event bad weather or disease prevents the farm from producing its expected yield. They also assume the responsibility for cooking and eating what is fresh and available in season.

At the same time, individuals who gave up gardening when they left home are beginning to return with a renewed appreciation for what they thought they had given up for good. Many of the people who buy seeds from me ask for specific varieties grown by their parents and grandparents. They are looking for taste, texture, and nutrition, as well as rekindling fond memories.

Small farm owners and operators are also realizing that turning over the entire genetic pool of not only fruits and vegetables but also soybeans, corn, and the cereal grains to large corporate farms is a serious mistake. Seed saving is making a comeback at all levels in the food chain, and many seed banks are being established. Terminator genes and soybeans with Brazil nut gene supplements have become as scary as tomatoes decorated with flounder genes.

The business model that assumed consumers would willingly accept a continuing decline in food quality has been shown to be faulty. As more and more individuals recognize the role of diet in maintaining good health, cheap food is seen as having too high a cost in terms of healthy living.

Going Back to the Future

Although gardening and small livestock raising went out of style after the 1950s, a lot of people are now realizing that gardening is good in more ways than one. When children participate in gardening, they experience nature in a very fundamental way. They see seeds sprout and watch as apparently dead parts of plants come back to life. The seed produces a vegetable, grain, root, or fruit plant, part of which can be eaten raw, cooked, canned, frozen, dried, or stored for future use. They experience true ripeness firsthand, even eating straight from the garden patch, rather than depending on tomatoes or strawberries grown in California and gassed en route to wherever they live. The flavors of homegrown fruits and vegetables cannot be imitated by those colored by ethylene gas (even if the gas is a natural one). Children can eat beans that have been allowed to reach full maturity on the vine, rather than those genetically modified for mechanical harvest and therefore picked long before the protein appears.

At my farmers' markets in Lexington and Berea, I often ask young children if they like tomatoes, and their answers are always brutally frank. A lot of them say no. When I ask why not, the explanation is usually the same: "They don't taste good." Embarrassed parents or grandparents always hasten to add that they are trying to get the youngsters to enjoy eating fruits and vegetables, but many children have been turned off by grocery store tomatoes that are hard as rocks with little or no discernible flavor.

Children who grow and harvest their own fruits and vegetables are much more likely to try them and like them, if for no other reason than pride in something they produced themselves. When they participate in cooking or canning or other food preservation methods, they learn the value of good foods that are eaten or preserved at the peak of flavor and nutrition.

Most schools, even those in depressed inner cities, have enough green spaces for school gardens. If part of the White House lawn can be turned into a vegetable garden, surely most schools can devote part of their grounds to gardening. Vacant lots in many cities have been turned into productive gardens, providing fresh fruits and vegetables to inner-city residents whose grandparents probably grew most of what they ate.

Although many towns and cities actively discourage gardening and the raising of chickens for eggs and meat (passing laws against such practices), some city leaders are rethinking that position. They rightly reason that well-kept gardens would not depress housing values, and laid-off workers could at least grow and preserve some of their own food. And getting out of the house and doing something useful can be a good antidote to feeling sorry for oneself.

Concurrent with the growing of school gardens, physical education needs to be reintroduced, along with recess. Children who are allowed to be active in an organized fashion in physical education classes use up a lot of energy (and calories). Recess times that encourage a multitude of running and jumping activities can free up bodies and minds that are too often numbed by a multitude of prescription (and other) drugs.

When I was a graduate student in physical education at the

University of Tennessee in Knoxville more than fifty years ago, my assistantship involved teaching classes in physical education in many of the city's elementary schools. The classroom teachers also took part, so they could improve their skills in teaching physical education. I could always count on the students being very excited when I rode up on my bicycle (my only means of transportation). As soon as I walked onto the playground, I would have a different youngster hanging on to each of my ten fingers. In one of the low-income sections of town, the students had to play on asphalt; there was no grass to be found. But whether inside or outside, all the students in my classes were very active, and at the end of their twenty to thirty minutes with me, they were panting for breath and ready to get back to work in the classroom. In the sixteen schools I worked in that year, I don't remember seeing a single student that I would consider overweight. And I never had a discipline problem with any student.

Two years later, after I had fulfilled a portion of my military obligation, I taught physical education in all eight grades at Pleasant Ridge Elementary School in Knox County, Tennessee; I also taught science and health to the eighth-graders. The school was in a transitional area: many of the students came from very low-income families; others were the children of Oak Ridge professionals who lived in the suburbs on the edge of Knoxville. Just like the children in inner-city Knoxville, the students at Pleasant Ridge were very active physically, and I stretched them to their limits. We played sports, including soccer; did tumbling and gymnastics; and practiced a lot of folk and square dancing. Some of the eighth-grade dancers even appeared on a Knoxville television program. I remember only one student in that school who was overweight, and it wasn't because of a lack of physical activity. My eighth-graders also got additional exercise when we took to the nearby hills for science class in the forest, gathering plants to create a terrarium and forest creatures to make it a home.

But in the name of progress, physical education and recess have virtually disappeared from many, if not most, of our schools, and prescription drugs have taken their place. A host of medical conditions, many with unintelligible acronyms, have been discovered that require many students to take daily doses of medication to stabilize their moods and their fidgetiness. One student who participated in a program I ran at Berea College in 1966 met me thirty years later and expressed thankfulness that Ritalin had not been invented when he was in high school. Both he and I knew he would have been a prime candidate for behavioral medications if they had been available back then. Instead, he used his abundant energies to become a highly competent individual in many fields and in his human relationships. He worked for IBM and was an outstanding musician, as well as a minister.

What much of this boils down to is that physical activity, including gardening, can go a long way toward healing. Working with the soil uses up calories just as surely as any gymnasium workout, and the foods produced through that labor can lead to healthier eating. It's like Robert Frost's idea that chopping wood warms you twice: first while chopping it, and then when it's burned in the fireplace or stove.

The United States has long had a cheap food policy, and laws passed by Congress have gone a long way toward making that possible. *Efficiency* has been the byword; *quality* and *health* are forgotten words. We have become so efficient that California, Florida, and a few midwestern states could provide cheap food for the entire country and much of the rest of the world.

Medicated animals fed heavy doses of corn, wheat, and other grains gain weight fast and end up on our tables in record time, too often with E. coli, salmonella, listeria, and other organisms in tow. Genetically modified plants are still highly susceptible to disease outbreaks, but they can provide uniform, machine-harvestable fruits and vegetables almost as fast as eighteen-wheel trucks can make a round-trip. High-fructose corn syrup saturates an increasing number of foods. And turkeys have to be produced through artificial insemination. We have a brave new world of animals designed just for the parts of them we like to eat most.

A record number of children are obese, and a record number are developing adult-onset diabetes. It is no longer unusual for people in their late twenties and thirties to have heart attacks. More and more people are having stomach-stapling operations and other procedures to limit their food intake. Sleep-inducing medicines are produced by the ton, and other medicines control our moods when awake.

In short, cheap foods and the attendant cultural manifestations of our consuming culture are proving to be increasingly expensive. Hospitals combat a host of disease organisms resistant to antibiotics. School cafeterias offer a bewildering assortment of heavily sweetened foods. Physical education programs are in danger of becoming extinct, and recess is a quaint reminder of times long past.

The "No Child Left Behind" program has morphed into an "All Children Left Behind" program when it comes to physical fitness and healthy eating habits. Teachers are required to "teach to the test," while the arts and other "nonquantifiable" activities are pushed aside to make way for more memorization and more testing. At the same time, studies have shown that physical education, the arts, and just plain free time are important in teaching children to "get it all together" into a meaningful whole.

I would propose gardening—at school, at home, and in the community, if space allows—as a way for children to get it all together. Time devoted to sowing, cultivating, weeding, reaping, cooking, eating, and relaxing is time well spent. Given our preoccupation with electronic gadgets and programmed activities, time spent communing with the natural world and producing at least some of our own food would be a healthy countermeasure to the pressures of modern society.

Of course, we can't blame all society's ills on our compulsion to have cheap foods. But starting again to have a more balanced relationship with our foods and food sources can go a long way toward restoring balance and civility in our lives, and especially in the lives of young children.

Introducing the next generation to seed saving. Bill Best at the Pikeville seed swap, 2015. (Dobree Adams)

Afterword

Collecting Colonial Heirlooms

I'm often bemused when enthusiasts debate how old a variety has to be to qualify as an heirloom. Probably the most often used figure is fifty years. Others chime in and insist that "antiques" are defined as being at least a hundred years old, so heirloom vegetables should use the same figure. Still others insist that because the shift to hybrids picked up momentum during and just after World War II, heirlooms should include only varieties grown previous to 1940.

My area of concern is southern colonial and federalist varieties. So while the argument rages, I quietly continue my search for varieties that are 250, 300, even 500 years old. Those, my friends, are heirlooms by anyone's definition.

A word of explanation may be in order. I came to the heirloom seeds movement about a quarter century ago, not so much as a gardener but as a food historian. My wife and I are living historians, reenacting the lifestyles of eighteenth-century America. Indeed, we are so into it that we serve as consultants to living history museums on the subjects of gardening, agriculture, and foodways. For instance, I put in the historical gardens at Fort Boonesborough State Park in Kentucky and managed them for several

Orinoco, the original tobacco variety grown by John Rolfe at Jamestown in 1610, still finds a home at living history museums such as Kentucky's Fort Boonesborough State Park. (Barbara Grant Elliott)

years. Along the way we have written two books about colonial cookery and produced a line of products utilizing herbs in the eighteenth-century manner.

Back around 1990 we decided to plant a kitchen garden using varieties from the colonial period. For example, we didn't want to plant just turnips; we wanted Dutch Yellow, which was a popular variety of the day. One of the first things we discovered was that Dutch Yellow, along with 80 percent of all eighteenth-century varieties, is extinct. Let me repeat that figure: a full 80 percent of eighteenth-century vegetable varieties are extinct. Seed for them doesn't exist anywhere in the world! Talk about agri-shock! I was appalled by that figure and took it very personally. "Who did that?" I exclaimed. "I want his name!"

The upshot is that I was thrust into the heirloom seeds movement, which was just picking up steam. For the next twenty-odd years, my dual interest in heirlooms and colonial foodways and agriculture developed side by side and continues today. And while I still can't grow Dutch Yellow turnips, I do grow Amber Globe, which resembles it in size and color and can be dated to before 1840.

One caveat: although much of what I have to share with you applies universally to British North America, my primary focus is on the region that would become the states of the mid-South: southern Maryland, Virginia, West Virginia, Kentucky, North Carolina, and Tennessee. Do not assume that what was true in this region applies equally in either the Deep South or the Northeast, which had their own vegetable types, varieties, and growing methods. In addition, my concentration only peripherally encompasses Native American foodways, which is a study in its own right. For starters, let's debunk a few myths.

Myth Number One

There was a limited number of vegetable choices on colonial tables. Nothing could be further from the truth. With the exception of a few weird choices (kiwi fruit, for example), virtually anything found on American tables today was available in the seventeenth and eighteenth centuries. Indeed, any vegetable type grown today could likely be found in colonial kitchen gardens and farm fields. What's more, they had many veggies that we don't grow much anymore because they've gone out of fashion for one reason or another. Cardoon, for example, which looks like celery on steroids, was very popular. Salsify—the so-called oyster plant—is another. And although we rarely grow fava beans, they were one of the more popular legumes in British North America, second only to English peas in importance.

Upper Ground Sweet Potato squash, originally from Tennessee, is popular in Kentucky and Ohio as well. This specimen came from Randy Wolf. (Barbara Grant Elliott)

Obviously, the mistress of a plantation, whose kitchen garden might encompass four acres, grew a greater diversity of plants than a pioneering housewife in Kentucky, whose kitchen garden was a small plot alongside the family's cabin. Nor can we neglect the unsung patches tended by slaves for their own use. In fact, many of the foodstuffs we think of as iconically southern—cowpeas, peanuts, okra, sweet potatoes, most chilies—were more likely to be found in one of the slave patches than in the kitchen garden itself.

A partial list of garden, field, and orchard plants commonly available in the colonial period would include the following:

Fruits: Apples, strawberries, blueberries, blackberries, gooseberries, cranberries, huckleberries, pears, peaches, plums, limes, oranges, lemons, quince, persimmons, grapes, and currants.

Grains: Maize (i.e., Indian corn), barley, wheat, buckwheat, oats, rye, and rice.

Roots: White (Irish) potatoes, sweet potatoes, yams, radishes, carrots, parsnips, beets, onions, garlic, salsify, and turnips.

Vegetables: Common beans, fava beans, English peas, cowpeas, pumpkins and other squashes, broccoli, cauliflower, cucumbers, celery, asparagus, mushrooms, wild and cultivated greens, sweet peppers, chilies, cabbage and all its relatives, lettuce, herbs, melons, and tomatoes.

Note that I list herbs as a vegetable. There was no particular differentiation in the eighteenth century; herbs were thought of as vegetables and were eaten that way—that is, served on their own—in addition to being used for medicinal, cosmetic, and flavoring purposes. Herbs were so important that basil was actually grown as a cash crop in parts of Virginia in the late 1700s. Herbs are the subject of myth number three, but another vegetable—the tomato—is the subject of the next myth and deserves a discussion of its own.

Myth Number Two

Neither the British nor British colonists ate tomatoes because they were thought to be poisonous. This one has some basis in fact. Tomatoes, a member of the nightshade family, were introduced to Europe in the 1500s by the Spanish, who had discovered them in South America. They spread rapidly on the Continent, and the Spanish, French, and Italians incorporated them into their cuisines. But the British were resistant; if they grew tomatoes at all, they did so as ornamentals. But there were exceptions, even in England. Hannah Glasse, in her 1745 *The Art of Cookery Made Plain and Easy,* provides at least one recipe using tomatoes.

The southern colonies, particularly Virginia and its satellites, were less resistant to tomatoes than the northern ones. As early as 1781, in a report he prepared for the French legate in Philadelphia, Thomas Jefferson identified tomatoes as a food-stuff. Jefferson did not grow them himself until the nineteenth century, but only because his political duties kept him away from home for so many years. He most certainly ate tomatoes while in Europe.

My favorite example, however, comes from William Whit-ley, who built the first round racetrack in America. Starting in 1788, included in his lavish race-day breakfasts was a dish called "stewed tomatoes." These were different from today's stewed tomatoes—the ones that come in a can and say "Hunts" on the label. Rather, stewed tomatoes was a complex dish that combined tomatoes, sugar, seasonings, butter, and stale biscuits to create a pudding-like product.

Here's a question: If Whitley was serving a sophisticated dish like this out in the Kentucky wilderness, is it reasonable to assume that people in Williamsburg and Richmond and Washington (with its large French population) weren't eating tomatoes? Of course they were. But this doesn't mean that tomatoes, or any vegetable, were necessarily used the way they are today. Yellow Pear cherry tomatoes, for instance, can be traced at least to 1750. But they weren't eaten out of hand or mixed in a salad, as we're likely to do. Instead, they were put up as preserves.

Colonial-era tomatoes were most likely to be yellow (rather than any of the other seven colors), with large shoulders and deep fluting. It wasn't until a hundred years later in the nineteenth century, when commercial canners pushed the development of the smooth, round, red standard, that things changed. Just as today's varieties are developed to fit the needs of the food distribution system, the smooth, round, red tomato was raised because of its ease of processing; it had no other intrinsic value over other forms.

Myth Number Three

Herbs, spices, and aromatic vegetables were used to cover the smell and taste of spoiled meat. This is perhaps the most enduring myth in the culinary world. This idea stems from the fact that people project backward when trying to interpret how early Americans ate. They start with an erroneous premise—that is, that early Americans ate bad food. Then they reason out an explanation and apply it as if it were fact. Living historians, in contrast, know that the worst thing you can do is apply a twenty-first-century mind-set to eighteenth-century lifestyles. The truth is, other than during times of privation, people living in the colonial and federalist eras were no more likely to eat spoiled meat than we are today. In fact, starting as early as the 1600s, cookbook authors often devoted much space to advising readers how to choose meats and vegetables to assure freshness and wholesomeness.

Until it moved in the 1980s, the oldest continually operating market in America was Boston's Haymarket Square. Produce and farm products were sold on that site as early as 1630, and by 1633, the colonial government had posted rules for vendors. Among them: "No blown, tainted, or spoilt meat may be offered." Severe penalties were imposed on violators.

What may contribute to this myth about herbs is the fact that people back then liked much bolder flavors than we do today. They preferred meat and poultry a bit on the high side and used herbs, spices, and other flavorings—particularly sours and bitters—with a heavy hand. Thus, many dishes eaten with relish back then would be considered inedible today. For instance, when making pickles nowadays, we start with 5 percent vinegar and often cut it with water, sugar, or both. In the eighteenth century vinegar was much stronger—upwards of 15 to 20 percent—and was used without dilution. Sometimes vinegar and verjuice (an extremely sour liquid made from unripe grapes or crab apples) were used in the same recipe.

To be sure, herbs were used to cover bad odors—but not in food. The practice of strewing herbs on the floor, for instance, was very common. When aromatics such as rosemary, lavender, hyssop, and sage were stepped on, they released their aroma, thus deodorizing the cabin and freshening the air. These same herbs were used to keep stored linen and clothing fresh and wholesome smelling.

Body washing was frowned upon in those days. Medically, it was believed that frequent bathing was unhealthy. So aromatics were used to mask body odors. Plum Granny, a melon-like fruit with an incredible aroma, was often used for that purpose and had the advantage of growing wild in American forests. The word "nosegay" stems from the floral and herbal accessories worn by women to mask their own body odors, as well as those in their environment.

The Difficulties of Collecting

One of the problems inherent in collecting colonial-era heirlooms is that their forms and uses have changed over time. And although there are exceptions—especially among the landed gentry, who exchanged plants, seeds, and other horticultural products—few references use varietal names. Within families, a certain tomato might be referred to as "that yellow one," or they might speak of the "beans with the big seeds" or call a particular type of chili the "mango pepper." Upscale collectors like Thomas Jefferson, who imported seed directly, might use an Italian or French name that, more often than not, had been assigned arbitrarily.

Hybridization also makes collecting difficult. Even before the eighteenth century, plant breeders were trying to create improved strains. This process picked up momentum in the late eighteenth century and continued from there—leading, inexorably, to the genetically modified organisms (GMOs) of today.

Take Bullnose peppers as a case in point. In the eighteenth

century the Bullnose was a small bell pepper, with no particular heat nor any particular sweetness. It was used to make mangos, which are peppers that are stuffed and pickled. (In some parts of America, such as Indiana, small peppers are still called mangos, although most people no longer know why.) Throughout the nineteenth century, Bullnose breeders selected for size and sweetness. By the end of that century, the Bullnose was a large, sweet pepper and the number-one commercially grown bell pepper in America. Although seed for the Bullnose is still readily available, it's not really the same pepper grown in colonial and federalist gardens. Far too often, as "improved" varieties are introduced, the original falls by the wayside. One of my many planned projects is to select backward and develop a strain of Bullnose that more closely resembles the original.

Name changes can also make collecting difficult. In the first American cookbook, published in 1794, Amelia Simmons refers to Clabbord beans. These pole beans were grown up the sides of buildings—in the clapboards, as it were. I spent several years trying to identify them, with no success. Then Wesley Greene, an agricultural historian at Williamsburg, came to my rescue. Clabbord beans are the same variety that Fearing Burr refers to as Case Knife beans in his seminal 1863 work *Field and Garden Vegetables of America*. Burr notes that they were also called Sabre or Cimeter. That, in itself, is a clue to its origins, as in the late seventeenth and early eighteenth centuries, these varieties were reimported to America as French Scimitar or Dutch Scimitar beans.

Colonial-Era Cookbooks

If you know what people ate, you have a pretty good idea what they grew, gathered, and hunted. So period cookbooks—of which there are a surprisingly large number—are a great place to start gathering information.

Fife Creek okra, a Native American variety, was gifted to the Fife family by a Creek Indian visitor in 1890. The Fifes have been growing it ever since in Mississippi and Tennessee. The seed for this Fife Creek okra plant at Fort Boonesborough came from a Fife family member. (Barbara Grant Elliott)

To be sure, gardening books were published as well. But with the notable exception of Thomas Jefferson's *Garden Book,* most of them deal with techniques and instructions for growing ornamentals, rather than vegetable gardening per se. What's more, they often assume the reader already knows what they are talking about, so much is left out. A perfect example comes from Robert Squibb's 1781 *The Gardener's Calendar,* in which he notes, "Your tomatoes will now begin to run: they, being of a procumbent growth, should have sticks to support them; which should not be very high, but strong and bushy." Is he referring to straight sticks, like those we use today? Or did eighteenth-century gardeners use sticks with many cross branches, such as the so-called pea sticks used for those legumes? From the word "bushy," we can presume he meant the latter. Other references, however, are vague at best. In McMahon's *The Gardener's Calendar of 1806,* he merely recommends that they (tomatoes) have sticks placed next to them.

Despite their usefulness, there are a couple of problems with cookbooks. First is nomenclature. Few of them use varietal names; they just refer to ingredients generically, such as beans, potatoes, pumpkins, and so forth. Within that framework we have the broader use of names. For example, the word "pompian" is an obvious reference to "pumpkin." Less obvious is that in colonial times, all winter squashes were called pumpkins.

In other cases, varieties are differentiated only by color. For example, when Amelia Simmons talks about carrots in *American Cookery,* she says, "the yellow are better than the orange or red." This doesn't help us identify the specific carrots grown in the latter part of the eighteenth century, but at least it tells us that three colors were available. Sometimes it's not that obvious. Ms. Simmons tells us that among watermelons, "the red cored are highest flavored." This lets us know that watermelons with other-colored flesh existed.

Don't be misled into thinking that no varietal names were used. They were. But they tended to change through the years. As with the Clabbord bean, a chain of names might be applied to the same vegetable. In other cases, the name remained the same—the Broad Windsor bean immediately comes to mind. Thus, for the collector, tracking down seeds often means tracing these name changes, and the *Oxford English Dictionary* can be a better reference than any gardening or cookery book ever written.

Another problem with cookbooks is their target audience. Cookbooks and cookery manuscripts were used by the literate members of the female population, who necessarily belonged to the upper class. As such, the recipes represent a more sophisticated level of cuisine and rarely refer to the foods and dishes of everyday folks.

Clues to how regular people ate can be found in journals, diaries, and letters. Many a journal entry reports, "dined on pumpkin and milk," or some variant thereof. Similarly, an early entry in the records kept by the Moravian missionaries in North Carolina says, "Our food has been largely pumpkin broth and mush, which has agreed with us very well." I have never seen a cookbook entry for pumpkin broth nor, for that matter, one for pumpkin and milk. But the constant reference to such dishes tells us that they were fairly common. In the case of the Moravians, we also learn that corn was a mainstay crop.

These sources can also provide insight into how things were done. Daniel Drake, in his memoir about growing up near Maysville, Kentucky, before the turn of the nineteenth century, gives us this gem: Watermelons, he tells us, were grown inside the cornfields to hide them from prying eyes and light-fingered neighbors. He also mentions that he always grew a small plot of wheat, for family use only. From this we learn that even in fairly populous areas, families maintained vegetable gardens that included corn, watermelons, and wheat. That the wheat was grown strictly for family use is more than an idle comment, because flour was a cash crop. Like tobacco, the quality of flour destined for market was strongly controlled; in Virginia, for example, it had to be produced in a certified mill and meet certain criteria. Such restrictions did not apply to crops people grew for their own use.

As we've seen, collecting seed from plants grown in colonial and federalist times has all the difficulties inherent in collecting any heirlooms, plus a few unique ones. But it can be done. During my last year running the historical gardens at Fort Boonesborough, I had thirty-six demonstration plots. Of those, nineteen (including both the corn and tobacco cash crops) could be traced to before 1800. Almost all the others could be dated prior to 1820.

True, researching and collecting those seeds wasn't easy. But it's one of the most rewarding things I've ever done.

Brook Elliott

Vinson Watts tomatoes growing in the high tunnel.

Resources

Arnow, Harriette Simpson. *Seedtime on the Cumberland*. Lexington: University Press of Kentucky, 1983.

Berry, Wendell. *The Unsettling of America: Culture & Agriculture*. Berkeley, CA: Counterpoint, 2015.

Berry, Wendell, and Norman Wirzba. *The Art of the Commonplace: The Agrarian Essays of Wendell Berry*. Berkeley, CA: Counterpoint, 2003.

Best, Bill. *Saving Seeds, Preserving Taste: Heirloom Seed Savers in Appalachia*. Athens: Ohio University Press, 2013.

Black, Katherine J. *Row by Row: Talking with Kentucky Gardeners*. Athens: Ohio University Press, 2015.

Buchanan, David. *Taste, Memory: Forgotten Foods, Lost Memories, and Why They Matter*. Foreword by Gary Nabhan. White River Junction, VT: Chelsea Green, 2012.

Caduto, Michael J., and Joseph Bruchac. *Keepers of Life: Discovering Plants through Native American Stories and Earth Activities for Children*. Golden, CO: Fulcrum Publishing, 1998.

———. *Native American Gardening: Stories, Projects, and Recipes for Families*. Golden, CO: Fulcrum Publishing, 1996.

Estabrook, Barry. *Tomatoland: How Modern Industrial Agriculture Destroyed Our Most Alluring Fruit*. Kansas City, MO: Andrews McMeel, 2011.

Gettle, Jeremiah. *The Heirloom Life Gardener*. New York: Hyperion, 2011.

Gift, Nancy. *A Weed by Any Other Name: The Virtues of a Messy Lawn, or Learning to Love the Plants We Don't Plant*. Boston: Beacon, 2009.

Goldman, Amy. *Heirloom Harvest: Modern Daguerreotypes of Historic Garden Treasures*. New York: Bloomsbury, 2015.

———. *The Heirloom Tomato: From Garden to Table*. New York: Bloomsbury, 2008.

Henderson, A. Gwynn. "Dispelling the Myth: Seventeenth and Eighteenth Century Indian Life in Kentucky." *Register of the Kentucky Historical Society* 90, no. 1 (1992): 1–25.

———. *Kentuckians before Boone*. Lexington: University Press of Kentucky, 1992.

Jabbour, Nikki. *Groundbreaking Food Gardens: 73 Plans that Will Change the Way You Grow Your Garden*. North Adams, MA: Storey Publishing, 2014.

Jackson, Wes. *Consulting the Genius of the Place: An Ecological Approach to a New Agriculture*. Berkeley, CA: Counterpoint, 2010.

Jeffery, Josie. *SEEDSWAP: The Gardener's Guide to Saving and Swapping Seeds*. Boulder, CO: Roost Books, 2014.

Kingsolver, Barbara, Camille Kingsolver, and Steven L. Hopp. *Animal, Vegetable, Miracle: A Year of Food Life*. New York: HarperCollins, 2007.

LeHouillier, Craig. *Epic Tomatoes: How to Select & Grow the Best Varieties of All Time*. North Adams, MA: Storey Publishing, 2015.

Lewis, R. Barry, ed. *Kentucky Archaeology*. Lexington: University Press of Kentucky, 1996.

Lundy, Ronni. *Sorghum's Savor*. Gainesville: University Press of Florida, 2015.

———, ed. *Cornbread Nation 3: Foods of the Mountain South*. Chapel Hill: University of North Carolina Press, 2005.

———. *In Praise of Tomatoes: Tasty Recipes, Garden Secrets, Legends, and Lore*. New York: Lark Books, 2004.

Lundy, Ronni, with Johnny Autry. *Victuals: An Appalachian Journey, with Recipes*. New York: Clarkson Potter, 2016.

Madison, Deborah. *Local Flavors: Cooking and Eating from America's Farmers' Markets*. New York: Barnes & Noble, 2008.

Male, Carolyn J. *100 Heirloom Tomatoes for the American Garden*. New York: Workman, 1999.

McKibben, Bill. *Deep Economy: The Wealth of Communities and the Durable Future*. New York: Henry Holt, 2008.

——. *The End of Nature.* New York: Ballantine, 2006.

Minnis, Paul E., ed. *People and Plants in Ancient Eastern North America.* Washington, DC: Smithsonian Books, 2003.

Nabham, Gary Paul, ed. *Renewing America's Food Traditions: Saving and Savoring the Continent's Most Endangered Foods.* Foreword by Deborah Madison. White River Junction, VT: Chelsea Green, 2008.

Pellegrini, Georgia. *Food Heroes: 16 Culinary Artisans Preserving Tradition.* New York: Stewart, Tabori, & Chang, 2010.

Pollan, Michael. *In Defense of Food: An Eater's Manifesto.* New York: Penguin, 2008.

——. *The Omnivore's Dilemma: A Natural History of Four Meals.* New York: Penguin, 2006.

Quillen, Rita Sims. *The Mad Farmer's Wife.* Huntsville: Texas Review Press, 2016.

Ray, Janisse. *The Seed Underground: A Growing Revolution to Save Food.* White River Junction, VT: Chelsea Green, 2012.

Still, James. *Jack and the Wonder Beans.* Lexington: University Press of Kentucky, 1996.

Veteto, James R., Gary Paul Nabham, Regina Fitzsimmons, Kanin Routson, and DeJa Walker. *Place-Based Foods of Appalachia: From Rarity to Community Restoration and Market Recovery.* Renewing Americas's Food Traditions (RAFT), 2011. www.raftalliance.org.

The following people may be contacted for further information about Kentucky heirloom seeds:

Bill Best
1033 Pilot Knob Cemetery Road
Berea, KY 40404
(859) 986-3204
bill_best@heirlooms.org
www.heirlooms.org

Dobree Adams
Riverbend Farm
PO Box 475
Frankfort, KY 40602
(502) 223-1858
rbgnomon@bellsouth.net

Frank Barnett
fbarnett@bellsouth.net

Brook Elliott
Historic Foodways
PO Box 519
Richmond, KY 40476
historicfoodways@hotmail.com

A. Gwynn Henderson
Kentucky Archaeological Survey
1020A Export Street
Lexington, KY 40506-9854
(859) 257-1944
aghend2@uky.edu
www.heritage.ky.gov/kas.htm

Susana Lein
Salamander Springs Farm
PO Box 354
Berea, KY 40403
(859) 893-3360
www.LocalHarvest.org/store/M5606

Julie and John Marushkin
Clark County Public Library
370 South Burns Avenue
Winchester, KY 40391
(859) 744-5661
clarkbooks@gmail.com

Roger H. Postley
Tomatoes, Etc.
(859) 278-4846
rpostley@aol.com

Tomatoes, staked in the field. (Dobree Adams)

Acknowledgments

First, I would like to recognize six generations of gardeners and seed savers, starting with my maternal grandmother, Kate Casper Sanford. She had been a widow for many years when I came along and was dependent on her gardens, cattle, and hogs for her living. But she managed and gave away a lot of food in the process. She is the grandmother I mentioned earlier who had a special garden across the road from her house; she invited hungry people to take what they needed. (This was during the Depression, and a lot of people needed that garden.) She was a superb cook and always "put a bug in my ear" to let me know where I could find the tea cakes she had just made before our Sunday visits and my tours of her very neat gardens.

Next, I would like to thank my mother, Margaret Sanford Best, who taught me to appreciate heirloom beans when I was barely old enough to pick the cornfield beans growing on the lower parts of the cornstalks while she picked those higher up.

My wife of fifty-four years, Irmgard, has worked countless hours for many years sorting seed beans to mail out to customers in all fifty states and at least a dozen countries around the world.

Our children, David, Barbara, and Michael, started growing heirlooms in the early 1970s when they were ages six through ten. They continued through high school and college, and Michael continues to this day as an agricultural economics professor at Tennessee Tech University.

Our grandchildren Jennifer, Christina, and Brian Best helped out on many occasions, and Brian has been working full-time during summers since his twelfth birthday and nearly full-time during recent years while he has been in college.

Our grandchildren Alexandra, Ashley, and Nolan Toti have helped out when visiting during the summer months.

Our grandchildren Sarah and Anita Best have helped with selling at farmers' markets and growing and saving seeds. Sarah is assisting a professor at Tennessee Tech University with DNA testing on all seeds in the Sustainable Mountain Agriculture Center bean collection to determine if we have duplicates.

And our great-grandchildren Peter, Greta, and Lilah Hess have started gardening and know how to string beans and remove seeds from Candy Roasters to dry for saving. Great-grandson Isaac Hess is ready to start soon, and great-granddaughter Lauren Whittemore was already harvesting blueberries and tommy toe tomatoes when she was two years old.

I would also like to thank the many members of my extended family who offer an entire table of bean dishes at our family reunions. I often enjoy a plateful of beans prepared different ways, along with cornbread and butter.

Hundreds of people have sent me their bean and tomato seeds to give me a start or have visited my farm to trade seeds or share their family seeds. I would like to thank all those who have welcomed me to their farms and gardens and shared with me varieties they have been working on for decades and bear their family names. This includes former Berea College and Upward Bound students who continue to grow family heirlooms.

I appreciate the many farmers' market customers in Lexington and Berea during the past forty-five years, especially those transplanted eastern Kentuckians who have shared their family heirloom seeds and stories from many counties.

Through these many years of growing, collecting, and sharing, several individuals have been especially important because of their knowledge of, interest in, and passion for heirloom seed saving:

Joyce Pinson of Pike County first bought heirloom seeds from me more than ten years ago and now sponsors a highly successful seed swap each year on the first Saturday in April.

Excellent gardeners Harvey and Ada Dicken of Clinton County introduced me to many family members and friends who excel in heirloom seed saving.

Kentucky seed savers Willard Wynn of Rockcastle County, J. B. Mullins of Breathitt County, and Zeke Dishman of Windy in Wayne County have been especially important to me and my work.

The following individuals played important roles in making this book a reality:

Laura Sutton, who coaxed me to write this book, and Ashley Runyon and Patrick O'Dowd of the University Press of Kentucky's acquisitions staff were a great help through the first stage of the publishing process.

Ronni Lundy and Gurney Norman enthusiastically recommended publication by the press after reading the initial manuscript.

Leila Salisbury, director of the press, and her staff graciously worked with us through the many steps from manuscript to finished book.

Chad Berry and Chris Green of Berea College made it possible for a section of photographs to appear in color.

Jonathan Greene was helpful throughout our putting the book together with his publishing and book design expertise.

And of course there would be no book at all without the specific contributions of the following friends and colleagues:

Acknowledgments

Gwynn Henderson's foreword adds a deep historical perspective about Kentucky's seed-saving peoples. I am grateful to her and to the three professional archaeologists who reviewed her contribution: Darlene Applegate, Western Kentucky University; Kris Gremillion, The Ohio State University; and Jack Rossen, Ithaca College. I also thank Gwynn for introducing me to archaeological fieldwork.

Brook Elliott's afterword provides an excellent history of colonial foodways before so many of our edible plants became extinct and tells us why it's important to discover, maintain, and nurture those that remain.

Frank Barnett's stories, eloquently written, remind us of the value of having kinfolks and friends in rural areas and of collecting their stories as well as their seeds.

Susana Lein advocates permaculture practices and demonstrates the importance of improving traditional varieties of vegetables by selecting for vigor and productivity.

Roger Postley shares his love of heirloom tomatoes and illustrates what can be done in a city backyard to keep them going.

Julie Maruskin demonstrates the vast potential of libraries and dedicated librarians in promoting heirloom vegetables and teaching seed-starting and seed-saving techniques.

Lola Choinski takes us on a gardening adventure in the 1940s and early 1950s, a time when eastern Kentucky was pretty much self-sufficient in terms of food production.

Bill Leach eloquently explains the traditions behind so much seed saving and culture in the mountains of Harlan County.

Melody Rose acquaints us with some heirlooms from western Kentucky and shares her enthusiasm for seed saving.

Gary Perkins is a longtime seed saver in eastern Kentucky who, along with Brook Elliott and Melody Rose, started the Appalachian Heirloom Seed Conservancy, which lasted for several years.

Their annual meetings were held in the barn at our farm.

Rudy Thomas pays a heartfelt tribute to his mother, an old-time seed saver in Clinton County.

Finally, I owe a special debt to photographer and fiber artist Dobree Adams, who farms and gardens along the Kentucky River north of Frankfort and whose photograph of my tomatoes graces the cover of this book. Dobree turned my notebooks of interviews and drafts of proposed chapters into a book worthy of publication. Because of our shared interests and her firsthand knowledge of working with the land, we have built a partnership that neither of us could have envisioned initially.

Bill Best surveying his crops, summer of 2009. (Dobree Adams)

About the Author

Although he has no academic degrees in agriculture, Bill Best has been farming all his life. He was the North Carolina corn-growing champion in 1951 as a fifteen-year-old sophomore and 4-H member. That record, exceeding the old one by fifteen bushels, lasted nineteen years!

After one college course in agriculture in 1954, Bill did not like the direction that field was taking and majored in biology and physical education instead. He graduated from Berea College in 1959, followed by an MS in physical education and modern dance at the University of Tennessee in 1962 and an EdD at the University of Massachusetts–Amherst in 1973 with a concentration in Appalachian studies.

From 1962 through 2002, he was a professor, coach, and administrator at Berea College. He taught swimming at all levels, literature in the general studies program, and alternating courses in mythology and aquatic art for many January short terms.

For eleven years, Bill was also farming outside Berea in nearby Jackson County while his family lived in dormitories on the college campus. They bought another farm outside Berea in 1972 and have lived on it since 1973.

Bill helped found the Lexington Farmers' Market in 1973. Back then he was the youngest member, and now he continues to participate as the oldest. He started the Berea Farmers' Market in 1974, ran it for more than thirty years, and is still an active participant. He won an award from that market in September 2015. That award, now called the Bill Best Local Foods Leader Award, is presented annually to a member of the Berea Farmers' Market. He was also presented the first Community Resilience Award by Sustainable Berea in the fall of 2015.

Bill Best has long been an advocate of sustainable agriculture, buying locally grown foods, eating healthy, and getting children involved in exercising and growing their own food. He believes that teachers and school personnel should be involved with school gardens.

Over the years, he has collected more than 700 heirloom bean varieties and, as director of the Sustainable Mountain Agriculture Center, has hosted an annual seed swap the first Saturday in October for the past fifteen years. This event has been attended by over 300 people from about a dozen states.

He has been active and has published many articles in the field of Appalachian studies. He has written ten books, including *Saving Seeds, Preserving Taste: Heirloom Seed Savers in Appalachia*, published by Ohio University Press in 2013.

A popular speaker, he gives many talks each year to college and university classes, gardening groups, and seed-saving organizations.

He and his wife of fifty-four years, Irmgard, continue to grow several acres of beans and tomatoes each year for farmers' markets and for their seed-selling operation. Michael Best, their youngest son and an agricultural economics professor at Tennessee Technological University, also has a lifelong interest in seed saving and will soon be taking over the activities of the Sustainable Mountain Agriculture Center.

Bill and Irmgard have three children, eight grandchildren, and five great-grandchildren.

Index